Climate Change, Gender Roles and Hierarchies

This book examines changing gender roles, relations and hierarchies in an ethnic minority community in Central Viet Nam. After decades of war, the community continued its self-sufficient way of life in this remote forested and mountainous region, but in recent years it has been forced to respond to severe climate threats combined with sudden and destabilizing socioeconomic and regulatory change.

Through the use of both qualitative (interview-based) and quantitative research methods, the book offers insights into the complex interactions between climate, regulatory and socioeconomic changes – including, paradoxically, the emergence of significant problems for both the community and the environment in the wake of policies designed to protect the natural environment. Facing greatly increased food and livelihood insecurity, the women and men of the community were pushed into the mainstream market economy without being fully prepared to participate in an economy that is still very new to them. These sudden transitions caused major shifts in gender roles and hierarchies, opening up new possibilities for women to increase their social status in a highly patriarchal context, but also at a cost for both women and men as women's burdens increased and men's traditional roles and livelihoods were lost. This book examines recent trends, including unanticipated changes and new possible policy-related approaches, and draws international comparisons with other ethnic minority, indigenous and remote communities facing similar complex forces of change.

This book will be of interest to scholars and postgraduate students of climate change, gender, environment, public policy and development studies.

Phuong Ha Pham has been working as a researcher and independent gender consultant for international organizations in Viet Nam and in the Greater Mekong Subregion, focusing primarily on women's empowerment, gender and ethnicity, gender and agriculture/rural development, and gender and climate change.

Donna L. Doane has been working in universities and research institutions in Thailand, the Philippines, India, Japan and the United States, focusing on issues related to the informal economy, low-income women's economic empowerment and security, gender and ethnicity, and indigenous knowledge and technology blending.

Routledge International Studies of Women and Place

Series Editors: Janet Henshall Momsen
University of California, Davis
Janice Monk
University of Arizona, USA

Who Will Mind the Baby?
Geographies of Childcare and Working Mothers
Edited by Kim England

Feminist Political Ecology
Global Issues and Local Experience
Edited by Dianne Rocheleau, Barbara Thomas-Slayter, and Esther Wangari

Women Divided
Gender, Religion and Politics in Northern Ireland
Rosemary Sales

Women's Lifeworlds
Women's Narratives on Shaping their Realities
Edited by Edith Sizoo

Gender, Planning and Human Rights
Edited by Tovi Fenster

Gender, Ethnicity and Place
Women and Identity in Guyana
Linda Peake and D. Alissa Trotz

Brokering Circular Labour Migration
A Mobile Ethnography of Migrant Care Workers' Journey to Switzerland
Huey Shy Chau

Climate Change, Gender Roles and Hierarchies
Socioeconomic Transformation in an Ethnic Minority Community in Viet Nam
Phuong Ha Pham and Donna L. Doane

For more information about this series, please visit: www.routledge.com/ Routledge-International-Studies-of-Women-and-Place/book-series/SE0406

Climate Change, Gender Roles and Hierarchies

Socioeconomic Transformation in an Ethnic Minority Community in Viet Nam

Phuong Ha Pham and Donna L. Doane

Routledge
Taylor & Francis Group

LONDON AND NEW YORK

First published 2021
by Routledge
2 Park Square, Milton Park, Abingdon, Oxon OX14 4RN

and by Routledge
52 Vanderbilt Avenue, New York, NY 10017

Routledge is an imprint of the Taylor & Francis Group, an informa business

British Library Cataloguing-in-Publication Data
A catalogue record for this book is available from the British Library

Library of Congress Cataloging-in-Publication Data
A catalog record for this book has been requested

ISBN: 978-1-138-59911-6 (hbk)
ISBN: 978-0-429-48592-3 (ebk)

Typeset in Times New Roman
by Apex CoVantage, LLC

Contents

Figures

Photos

Tables

Preface

This study of the impact of climate change, together with sudden regulatory policy and rapid socioeconomic change, on gender roles, relations and hierarchies in an ethnic minority community in Viet Nam grew out of the doctoral dissertation of the lead author, Phuong Pham. One of her thesis advisors, Donna L. Doane, joined her in the journey toward expanding the thesis in terms of international comparisons, further analysis and additional fieldwork. The benefit of this approach has been to provide a longer time frame – with visits in 2011–2012, 2014 and 2019 – allowing a deeper understanding of the dramatic changes that have taken place in this commune.

In spite of the progress made toward understanding these changes, we have recently been struggling with the idea of the Co Tu ethnic minority of Ca Dy Commune in Quang Nam Province as being "relatively isolated" prior to the fundamental and destabilizing events of recent decades. How could we call the people of this commune "relatively isolated" prior to these recent developments when, in fact, they had been deeply affected by the French and Vietnam-American wars (the First and Second Indochina Wars) and could in no way be viewed as an isolated community with few outside contacts?

Interviews with older members of the community, who often describe their community as "previously isolated and self-sufficient" (prior to recent changes), revealed that many of the Co Tu villagers living in the commune had, in fact, fought against the French after 1945, and one elder estimated that about 40% of the commune later supported the Northern Army during the American War (for reasons discussed briefly in Chapter 2). They did this by carrying food and weapons, with some joining the army in fighting. Moreover, because the commune was close to the Ho Chi Minh Trail, it saw some of the worst effects of the war. As one elder noted:

> *Terrible damage was caused to the forests and water sources. Fields were abandoned because of Agent Orange. During and after this period, many people left their villages because of Agent Orange and famines. . . . I was lucky when I found my countrymen [from the same village]. We formed a group and went back to our village together.*

His account was supported by others who reported that during this period villagers left the commune and hid in neighboring forests and mountains for many years and then returned after the war ended. They all stated that during the war the area had been destroyed by bombs, identified as napalm bombs, and Agent Orange (including the highly toxic chemical contaminant dioxin):

> *Houses and fields had been destroyed, and there was famine and diseases. Many people died. Everything was destroyed. Cassava somehow survived and saved the people. Many people – about fifteen in Ca Dy Commune – have been affected by dioxin. Among these, three are receiving government subsidies.*

In spite of these accounts, the older generation's experiences before 1975 were rarely brought up in conversations about changes the commune had experienced, possibly because the effects of the war years were no longer visible and subsequent generations had other pressing concerns. In addition, although their homes and possessions were burned and their natural environment destroyed during these years, only three of eight Co Tu villages in the area at that time (those closer to the lowlands) were forcibly relocated into some form of "strategic hamlets," which may have allowed the majority of Co Tu people to maintain their separate culture and way of life in spite of these extraordinarily difficult experiences.

Still, we wondered, how could the Co Tu of Ca Dy Commune be considered as "relatively isolated" before recent decades, when the older generation had lived through such times? And why did those interviewed see the major changes that had come to the community as occurring not during the war years but rather during *later years*, specifically the period beginning roughly 15 years after the war ended (i.e., the early 1990s), which is the point at which the present study begins its narrative and analysis of change?

The answer may lie in the fact that the ethnic minority residents of Ca Dy Commune were able to continue what they viewed as their traditional way of life, which was a virtually self-sufficient way of living in a remote mountain area that had minimal contact with the Kinh majority ethnic group or the mainstream economy and society until relatively recently. Elders stated that when people had to leave their villages during the war years, they did have interactions with the Kinh majority, including through some purchasing activities. They also mentioned that after villagers returned to the commune and rebuilt their homes and settled back into their lives, more Kinh came to the area and limited commercial transactions did take place between the mountain and lowland regions. However, their perspective was that theirs was a way of life that had always been self-sufficient, entirely dependent on nature, had its own cultural traditions (including beliefs and values that were central to their existence), was based on social solidarity and was remote enough in this mountainous region as to be virtually isolated from mainstream society. For this reason, they see the most significant changes as having come into the commune in very recent years.

The present study has thus focused on the complex interaction of environmental, regulatory policy and socioeconomic (including cultural) changes of the last two to three decades that have had major impacts on what we describe here as a "previously remote and self-sufficient ethnic minority community in a mountainous part of Central Viet Nam, one that was relatively isolated until recent years." This is consistent with how those in the Co Tu community in Ca Dy Commune viewed themselves, at least until climate threats and other events of recent years made a continuation of that earlier way of life impossible, and it is understandable why they see their history in this way.

This, then, is a story of how a convergence of changes upended the traditional social structures, particularly with regard to the *gender-related roles, relations and hierarchies* that had prevailed in previous years, as the ethnic minority community was forced to respond to climate change and other compounding forces, with both positive and negative impacts on the women and men of the community. We hope we have been able to convey their experiences and views accurately in order to contribute to a greater understanding of this ethnic minority community, and other indigenous communities within and outside the region, as they continue to face major ongoing environmental and socioeconomic challenges and rapid change, both now and into the future.

Acknowledgements

Many in Viet Nam, Thailand and other countries contributed their time, dedication and support, which made this study possible. Phuong Pham would like to acknowledge the following people, in particular:

I wish to express my sincere gratitude to my advisor at the Asian Institute of Technology, Dr. Philippe Doneys, for his thorough guidance during the entire process of my PhD and his encouragement for me to write this book; to Dr. Bernadette P. Resurreccion, who instructed me and pushed me to work hard during the first three years of my PhD program; and to Dr. Edsel E. Sajor and Dr. Joyee S. Chatterjee, my committee members, for their constructive comments and suggestions. I am also very grateful to Dr. Janet Momsen for encouraging me to write this book based on my research for my PhD thesis. I also am grateful to my former bosses and colleagues in the Environment Operations Center/Asian Development Bank, particularly for their support in my fieldwork and data collection. I want to take a moment to express my deep thanks to my dear friend and colleague, Dinh Minh Hieu, who provided invaluable support for my research work, and to my beloved husband for his support in different aspects over the years, and above all, for his quiet and sincere encouragement. I also wish to express my sincere thanks to Mr. Dao Trung Kien and Ms. Pham Hoang Giang from the QA Global Quantitative Research Center, who provided great help in strengthening the quantitative analysis for Chapter 2.

I am tremendously grateful to the Quang Nam Provincial Agriculture Department, the Women's Union of Nam Giang District, the Women's Union of Ca Dy Commune, and above all to the local people in Ca Dy Commune, especially Ms. Alang Thi Nhung, Mrs. Trinh Thi Hong, and Ms. Alang Thi Tam, for their support and help during the fieldwork that provided the foundation for this research. My special thanks go particularly to the households – women and men of all ages – who accepted my request to be involved in the process of data collection. This research would not be in any way possible without their patience and tolerance of all my questions and the thoughtful insights they offered during the interviews.

We both would also like to thank a number of individuals who helped in different stages of the course of preparing this manuscript, including Ruth Anderson, two anonymous readers of the original proposal for the book, Vo Trong Hoang, Linh Ta, Jiao Xi, Kerrissa Vaughn, Lothar Linde, Eleanor A. Winberg, Fazle

Karim, Mahbooba Amin, Ibipo Johnston-Anumonwo, the Tagawa and Dhar families, K. Jadet, Nonita Saha, Shubham Pathak and, of course, Javed H. Mir. They have all contributed significantly to making this book possible, and, in fact, we would not have been able to complete this without their assistance.

Finally, we would like to express our thanks to our families – and especially the "older" (parents, siblings and husbands and their families) and younger generations – who gave us time, space and assistance as we often worked too long into the night to complete this project. With this in mind, we would like to dedicate this book to our children, Khoi Nguyen and Hong Anh (Phuong Pham) and Rehan and Adnan (Donna Doane). They, and all the others noted, know how important they are to the completion of this book. For all they have done and for being as they are, we are deeply grateful.

Abbreviations

ADB	Asian Development Bank
CEMA	Committee on Ethnic Minority Affairs (Viet Nam)
COP	Conference of the Parties (associated with the UNFCCC)
CPC	Commune People's Committee (Viet Nam)
CV	Coefficient of Variation
FAO	Food and Agriculture Organization (UN)
FGD	Focus Group Discussion
FLA	Forest Land Allocation Program (Viet Nam)
Ha	Hectare
IAITPTF	International Alliance of Indigenous and Tribal Peoples of Tropical Forests
IDI	In-depth Interview
IFAD	International Fund for Agricultural Development (UN)
ILC	International Law Commission (UN)
IPCC	Intergovernmental Panel on Climate Change (UN)
ISPONRE	Institute of Strategy and Policy on Natural Resources and Environment
KII	Key Informant Interview
MK test	Mann-Kendall test
MONRE	Ministry of Natural Resources and Environment (Viet Nam)
NGO	Non-Governmental Organization
NTFP	Non-Timber Forest Product
NTP	National Target Program (Viet Nam)
NTP-NRD	National Target Program on New Rural Development (Viet Nam)
UNDP	United Nations Development Program
UNEP	UN Environment Program
UNFCCC	United Nations Framework Convention on Climate Change (UN)
UNICEF	United Nations Children's Fund
USD	United States Dollar (national currency)
VND	Viet Nam Dong (national currency)
WCED	World Commission on Environment and Development (UN)
WEDO	Women's Environment and Development Organization
WFP	World Food Program (UN)
WHO	World Health Organization (UN)

1 Introduction

Gender and climate change in a previously self-sufficient ethnic minority community

This study is concerned with gender and climate change, specifically within the context of a previously remote and self-sufficient ethnic minority community in a mountainous part of Central Viet Nam that is now facing serious challenges. It not only examines the destabilizing impacts of a changing climate, it also explores the complex interactions of climate change combined with rapid socioeconomic and regulatory policy changes – including, paradoxically, significant problems that have emerged for both the community and the environment as a consequence of policies designed, in part, to protect the forests and the natural environment. As a result of these changes, the Co Tu community in Ca Dy Commune of Central Viet Nam is experiencing food and livelihood insecurity and is being pulled into the mainstream cash-based market economy, even though they are in many ways not fully prepared to participate in an economy that is still very new to them.

All of these rapid changes have had major impacts on gender roles in the region, coming in the wake of the elimination of men's traditional livelihood activities and, as a result, the introduction of heavy new burdens placed on women to ensure their family's survival. The study has parallels in other parts of the world where previously remote and self-sufficient communities are suddenly disrupted by a number of simultaneous changes, including the crucially important impacts of climate change and other changes that render earlier agricultural and forest-based livelihood systems no longer viable as a means to sustain the local communities.

This study, using both qualitative interview and quantitative household survey research methods, grew out of the lead author's research for her PhD thesis and is supplemented by additional research visits, further analyses, and international comparisons with other similar cases.[1] The present expanded study will detail the impacts of these rapid transitions on gender roles, relations and hierarchies in this ethnic minority community.

It will be argued that even though these changes have created serious new problems, they have also opened up new possibilities for women to increase their social status; however, the women do this through overwork as they earn and support their families by taking on both the livelihood and reproductive burdens of the family and community. In contrast, the initial response of a majority of men in the community was idleness – contributing very little while older and younger women worked – given that their traditional roles were eliminated in this new context. This has parallels in many countries where men's earlier livelihood activities are no longer available (e.g., by being prohibited or made obsolete), and men find it difficult to move away from their previous roles and positions in society. The gender-related changes analyzed here go further and examine the additional impacts on younger generations, who now have to confront a new, difficult and challenging world unknown to the older generations in the community – again with very different responses based on gender.

The study thus aims to draw lessons that will be relevant to contexts where climate change is having a major destabilizing impact, and together with other factors is forcing previously remote, poor and isolated communities into a new and unequal relationship with the mainstream cash-based economy. This can cause major shifts in gender roles and hierarchies, as both women and men attempt to cope – and not always successfully – with their new physical, socioeconomic and interpersonal environments.

Background to the research

Viet Nam is ranked ninth among the ten countries most affected by climate change from 1998 to 2017, according to the Global Climate Index 2017 (Eckstein et al., 2018). With 3,200 km of coastline and 70% of its population living in low-lying deltas and coastal areas, the country is highly vulnerable to natural hazards. Of the country's population, 10.8% will be severely impacted by one meter of sea level rise – the highest in the region (Dasgupta et al., 2007). It is predicted that climate change will increase sea levels and add significantly to damage caused by natural hazards. As an example of Viet Nam's vulnerability to climate-related disasters, in 2017 persistent weather extremes, including Tropical Storms Kirogi and Talas and Typhoon Damrey, caused hundreds of deaths, destroyed thousands of houses and severely damaged infrastructure and water supplies (Eckstein et al., 2018).

These climate-related changes already have had major impacts on virtually all sectors of the economy, particularly on agricultural and forest-based production systems. Of note is the serious negative impact on food production from continuously increasing temperatures and significant rainfall variability in Viet Nam. For example, in late 2004 and early 2005, it is estimated that 60 million USD in agricultural production was lost due to a sharp reduction in rainfall, and 1.3 million people did not have access to clean water (Tran, 2009). A 2009 study conducted in Ninh Thuan Province by Oxfam and UN-Vietnam found that the severe drought conditions caused land degradation, a lack of drinking water and serious reductions in crop yields; more recently, conflicts resulting from drought conditions have also been documented in Central Viet Nam (Van Huynh et al., 2019).

By 2070, the rice crop, in particular, is projected to decrease 3–6% as compared to the 1960–1998 period. In a developing country such as Viet Nam, agricultural production is crucial to food security for the entire nation, and the impact of climate change is even more critical than is true for countries that are less dependent on agricultural production (Kryspin-Watson et al., 2006; Tran, 2009; UN-Vietnam, 2009). The concern is that the pressures caused by food insecurity as a result of climate change will undermine the efforts to erase poverty – not only in Viet Nam but in all such affected countries.

More than any other group in the country, the poor and marginalized communities living in rural areas are believed to be the most vulnerable to climate change because they remain highly dependent on local natural resources for food security and for their livelihoods. Furthermore, they are said to have the least capacity to adapt as a result of their poverty and isolation (Abeygunawardena et al., 2009; Dankelman et al., 2008; Parry et al., 2007).

In addition to climate-change impacts associated with poverty and marginalization, climate change is widely recognized internationally as having important gender implications, and this is as true in Viet Nam as it is elsewhere. Women are very often found to be among the most adversely affected by climate change.

Women in Viet Nam are disadvantaged in many aspects: They are more likely than others to be subject to poor economic conditions, have less access to education and suffer from a lack of opportunities to become involved in public events and decision-making processes because of social constraints and stereotypes held by a society that remains influenced by Confucian and related ideologies (Tran et al., 2006). In addition, women in Viet Nam – who account for two-thirds of the rural labor force – play a crucial role in subsistence agriculture. In many developing countries, including Viet Nam, women have traditionally been required to take on the responsibility for providing daily meals and other basic necessities for their families. For this reason, rural women are very vulnerable to climate threats because of their designated responsibilities as well as the nature of their livelihoods, which depend heavily on natural resources (Chaudhry & Ruysschaert, 2007; Dankelman et al., 2008; Oxfam & UN-Vietnam, 2009; UN-Vietnam, 2009). It is widely recognized that rural women are among the groups most vulnerable to climate change due to a combination of poverty, food insecurity and gender inequality (Brody et al., 2008; Lambrou & Nelson, 2010; WEDO, 2007).

It is also increasingly recognized that the impacts of climate change are shaped by local institutional and socioeconomic contexts (Chaudhry & Ruysschaert, 2007; McElwee, 2010). In Viet Nam, ethnic minorities often have very limited cash income and are much more dependent on natural resources than the Kinh population, the ethnic majority group.[2] With the exception of the Khmer and the Cham, who are settled in the Mekong Delta and the Southeast Coast, respectively, most of the ethnic minority population lives in the more mountainous and forested parts of Viet Nam (Swinkels & Turk, 2006).

For these forest-based and upland ethnic minority communities, even minor changes in their forest and agricultural systems have serious impacts on food security. Here, both direct impacts of specific climate events (e.g., floods, droughts and storms) and indirect impacts of climatic variation – including, for example, negative impacts of extreme temperatures or rainfall on vegetation, wild game, fish and livestock, along with other environmental impacts – would make them vulnerable to increases in both food and livelihood insecurity. Climate threats, therefore, can have devastating impacts on local peoples' lives and wellbeing.

In Viet Nam, climate change–related studies have focused primarily on large-scale events affecting coastal communities (e.g., unprecedented rises in the sea level) as well as those living in flood-prone parts of the country. Nonetheless, climate change has also had profound impacts on other communities, including those living in mountainous areas, and we find that it is in these fragile mountain regions that not only the environmental impacts but also the social impacts of climate change need to be studied in much greater detail. This would include impacts on *gender roles and responsibilities* in the wake of severe climate variations and destructive and ongoing climate threats. In Viet Nam, this is particularly important given the central role gender plays in determining *food and livelihood security* in vulnerable communities – in other words, in determining the means of survival for these local populations.

Focus of the research

Ca Dy Commune is located in a forested mountainous area about 70 km from the center of Quang Nam Province and about 10 km from the nearest town of Thanh My in Central Viet Nam (see Figure 1.1, showing the province, district and the specific commune that is the focus of this study). Of the total population in this commune, 86% is made up of the Co Tu ethnic minority group.[3]

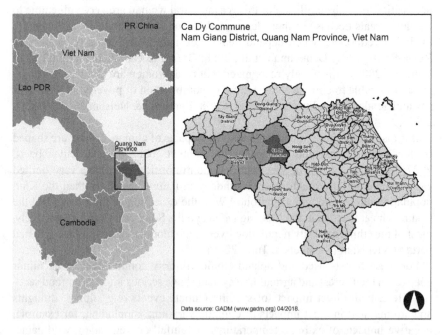

Figure 1.1 Study area – Ca Dy Commune, Nam Giang District, Quang Nam Province

Source: GADM (www.gadm.org) 04/2018

Traditionally, the livelihoods of the Co Tu community in Ca Dy Commune were based primarily on forest resources, including the collection of non-timber forest products (NTFPs), swidden (also known as slash and burn, or shifting) cultivation, the cutting and gathering of trees for houses, hunting, small-scale animal raising, and upland agriculture. This area was previously considered to be a remote and isolated part of Viet Nam, but changes have come due in part to the construction of the National Route 14B (a part of the Ho Chi Minh Highway) in 2004–2005, when the road began crossing through the commune and linking it with other parts of the country.

A study from 2011 analyzing the relationship between climate variations and land use in Quang Nam Province found that average air temperatures have increased about 0.06–0.1°C per decade from 1980 to 2010 (Vo, 2012). From 1991 to 2010, monthly rainfall has also been increasing with a standard deviation of 500–700 mm in most areas in the province. Due to the intensifying variability of rainfall, the frequency of floods and droughts in this province has also increased in recent years. In line with this, the Co Tu population living in this province has been experiencing significant climatic variations in recent decades, specifically regarding temperatures, rainfall and climate extremes. As a result, they find that their agricultural livelihoods have been under serious threat from climate change, as will be discussed in detail in the following chapters.

Apart from climate change, the Co Tu community in this area has also experienced major impacts brought about by changes in the land use and forest-protection laws and policies. The 1986 Doi Moi reforms initiated the implementation of a series of regulatory changes (i.e., changes pertaining to rules, laws and policies) throughout the country. The study area has been affected in particular by the 1991 Law on Forest Protection and Development and the 1998 Forest Land Allocation (FLA) program that allocated forestland to individual households, among other forest protection and forest management-related programs.

Due to these regulatory changes, the local community no longer has free access to the natural forests they depended on for generations. As a result, swidden cultivation – the most critical traditional livelihood of Co Tu women – was banned, and on top of this serious development, hunting – the most important livelihood activity of Co Tu men – has been strictly prohibited. The gathering of NTFPs has also been limited due to these restrictions. As a consequence of a changing climate in the context of a changing regulatory and social environment, the community's access to food and livelihood security has declined, forcing Co Tu women and men to cope and adapt as a means of survival.

In a world with a changing climate, harmonious relations between men and women can play an important role in enabling coordinated efforts and adaptation strategies. However, in this case, under pressure from the need to respond rapidly to a destabilizing climate and other changes, the established patterns of gender relations in the Co Tu community have been disrupted. New patterns of gender relations and new challenges to gender hierarchies are emerging and not always in a smooth and harmonious way.

Gender in a context of climate, regulatory and social change: a
survey of relevant literature

As noted, the focus of this research is an examination of changing gender relations
in an ethnic minority community under the serious impacts of a changing climate
in the context of regulatory and socioeconomic change. Climate change by itself
is already a major challenge, and its impact is intensified when compounded by
vulnerability factors, such as poverty, inequality and marginalization.

The following discussion will present a rather lengthy overview of related
studies, with the intention of exploring theories and empirical research that help
clarify these key research concerns. We should note that *readers already familiar*
with literature on many of these topics could safely move on to other topics they
know less about (i.e., reading selectively), or proceed directly to sections under
each topic that deal specifically with the context of Viet Nam.

Regarding the order of topics included in this overview, we will first discuss
the linkages between gender and climate change and examine how a consideration of gender has been incorporated into climate change–related discourse. The
next section will discuss livelihood and food security – two of the most important issues for rural and poor communities, such as those that are the focus of
this study – specifically from a gender perspective. Studies on men, masculinities
and work will then be presented to examine how existing research presents the
impacts of a changing society on men and conceptions of masculinity as well as
on gender relations. Next, the literature on climate change adaptation from a gender perspective will be reviewed briefly in the fourth section as another key body
of writings that inform this study. Following this, given that the community under
study is an ethnic minority community, the fifth section will focus on the vulnerability of indigenous and ethnic minority populations to climate-change impacts,
again with a specific concern for gender implications. Finally, relevant literature
on gender and climate change, particularly with regard to climate-related policies,
will be reviewed in the context of Viet Nam.

Although the number of topics covered make this review unusually lengthy
(and thus readers are encouraged to read selectively), all of these perspectives
have helped inform the current study, and we will return to the findings of this
research in relation to the relevant literature in the last chapter of this book.

Representation of gender in climate-change debates

It has been globally well recognized that rural women are vulnerable to climate-
related changes (Adger, 1999; Altieri & Koohafkan, 2008). They are also
considered key active agents of climate-change adaptation due to their gender-
differentiated roles, but in many country contexts – including in Viet Nam –
women face multiple obstacles in adapting to climate change in comparison with
men. In such contexts, women usually have fewer assets and limited control over
resources and consequently have less adaptive capacity than men. According to
Colfer et al. (2016), their ability to act on information concerning climate risks

and adaptation measures is constrained in a majority of developing countries by their lack of education and limited access to information, their dependence on natural resources, a gender-unequal division of labor, limited mobility and their limited role in decision-making.

Given that much of the literature on gender and climate change tended initially to focus separately on either women or men, the concern of more recent writers is that many studies end up deemphasizing the ways in which the effects of climate change may act to either challenge or reinforce gender relations and gender hierarchies (Buckingham & Le Masson, 2017). Gender has been increasingly addressed in climate change–related discussions, but these discussions often end up being focused on women's vulnerabilities and dichotomize men and women as homogeneous categories or limit themselves to comparing male- and female-headed households (Van Aelst & Holvoet, 2016). This type of analysis can be useful as well, but the unequal processes of decision-making that contribute to this vulnerability are often hidden in these discussions (Arora-Jonsson, 2011). Moreover, recent studies using an intersectionality approach also suggest that gender dynamics are complicated and are influenced by age, ethnicity, income status, and other factors (Carr & Thompson, 2014; Tschakert, 2014). This approach is relevant to the context of Viet Nam and is thus employed in the current study.

As noted earlier, much of the discussion regarding gender and the environment began with a concern for women as a general category and not with different contexts and types of gender relations in a more nuanced way. It is widely accepted that the more disadvantaged people are, the more vulnerable they are to climate change. Levels of infrastructure, economic development, social inequality and political influence of the affected countries and communities are among the key factors determining the extent of a local population's vulnerability to climate threats and environmental degradation (Masika, 2002). Among these, poverty is one of the most important variables affecting one's ability to achieve adequate self-protection. The quality and location of housing; adequate storage for food; access to healthcare services; productive resources, such as income and employment; and other key factors are crucial to adaptive capacity. Those who live in areas with the highest risk are often those with the least social and economic power and also have the least ability to cope with, and recover from, disasters associated with climate change.

In developing country contexts characterized by the feminization of poverty, women are generally acknowledged as being more vulnerable than men to the effects of climate threats. Poverty and unequal power relations are the main reasons why rural women in developing countries are often found to be the most adversely affected by negative impacts of climate change (Denton, 2000). Moreover, cultural norms can amplify the problems caused by poverty and unequal power relations when facing climate-related disasters (Alston, 2015). Women are likely to experience high levels of "pre-disaster" poverty because their responsibilities in the domestic sphere make them economically vulnerable before such an event even occurs; thus, a gender-differentiated division of labor – along with gender-differentiated access to income and productive resources that can provide security, protection and recovery – are causes of women's greater vulnerability to climate change and natural disasters (Cannon, 2002; Denton, 2002; Masika, 2002). In addition to large-scale disasters, less-serious weather events, such as heavy rain or dry spells, can also add serious impediments to women's daily roles and tasks, thereby increasing their burdens (Dankelman, 2010). At the same time, due to their roles in both productive and reproductive spheres, women play a key part in climate-change adaptation and mitigation (Alston, 2015), as will be discussed later in this chapter.

It is clear that the literature on gender in climate-change debates has been very insightful in terms of gender-differentiated *impacts* of climate change. One of the challenges now is to focus additionally on gender relations and hierarchies and how these may affect women's and men's very different *responses* to climate change, including bringing new and more gender analyses into climate science and policy debates (Alston & Whittenbury, 2013). Moreover, as noted previously, it has proved centrally important to move away from discussing women or men as general categories and to move toward a more nuanced understanding of contexts faced by different groups of women and men according to their age group, ethnicity, income status and other key contributing factors. This approach has been used extensively in the present study.

Gender, livelihood security, food security and climate change

Livelihood can be defined in a general way as a means of making a living to secure the necessities of life. The concept of *sustainable livelihood security* was proposed in a report by an advisory panel of the World Commission on Environment and Development (WCED). It suggests that a household can be said to have a sustainable livelihood if it has adequate stocks and flows of food and cash to meet basic needs on a long-term basis. A sustainable livelihood can be said to be secure if the household has ownership of, or has gained access to, resources and income-earning activities (WCED, 1987); more specifically, a livelihood can be considered sustainable if it can cope with and recover from stress and shocks, maintain or enhance its capabilities and assets and provide sustainable livelihood opportunities for the next generation (Chambers & Conway, 1992; Hussein & Nelson, 1998).

As highlighted earlier, impacts of climate change are disproportionate depending on specific socioeconomic factors, including poverty or wealth. Climate-change impacts on the poor and other vulnerable groups can in part be understood through the dynamic of negative impacts on their livelihoods, making them unviable or unsustainable, which causes further impoverishment. "Poverty is one of the main aspects of vulnerability but it varies with occupation and social characteristics, such as gender, age, ethnicity and disability" (Momsen, 2010, p. 130).

As climate change results in a negative impact on the *livelihood security* of vulnerable groups, it may equally have a profound impact on their *food security*. Along with population growth and socioeconomic issues, climate change is identified as one of the main threats to food security. As an example, droughts were reported by the FAO to cause more than 80% of the total damage and losses in agriculture during the ten-year period of 2006–2016, especially for livestock and crop production (FAO, 2017). In addition, forest-based livelihoods and food sources and reserves are seriously threatened by storms, floods, droughts, fires and infestations associated with climate change.

Despite efforts throughout the world to address food insecurity, hunger is on the rise. The number of people facing "crisis-level food insecurity" is said to have increased from 80 million in 2015 to 108 million in 2016 and 124 million in 2017 (FAO et al., 2018). The 2030 Agenda for Sustainable Development, therefore, calls for interventions designed to build resilience and adaptive capacity through short-, medium- and long-term policies, programs and practices in response to natural hazards and climate-related disasters in all countries, in alignment with such global policy platforms as the 2015 Paris Agreement regarding climate-change adaptation and mitigation and the Sendai Framework for Disaster Risk Reduction.

Poor rural women in developing countries are identified as one of the groups most vulnerable to both livelihood and food insecurity caused by climate change, given that their livelihood activities and their designated gender roles depend directly on the natural environment. Particularly with the increasing migration of men to urban areas, the literature emphasizes that women often remain in rural areas and take the lead in agriculture production and care work. With women comprising up to 60% of the rural poor in developing countries, this double burden increases their vulnerability to all of the negative impacts of climate change (United Nations Environment Program, 2016). For example, Oxfam (2007) found that in a community in Ninh Thuan Province of Central Viet Nam, women bore chief responsibility for collecting water for home use, and, as a result, they were severely affected by intensifying climate-related droughts. Then, in 2016, Viet Nam experienced the worst drought in 20 years, caused by the El Niño phenomenon (FAO, 2016). In the Mekong Delta and South Central regions, 75% of households reported a 35–38% reduction in household incomes. During the peak of the drought, an estimated 2 million people did not have access to water for consumption and domestic use, 1.1 million were food insecure, and more than 1.75 million people lost incomes due to partially or totally lost livelihoods. This has affected women in particular, for example, by increasing the average time they spent collecting water by 2–3 hours, thus increasing their work burden and negatively impacting the time they could spend on livelihood activities.

These studies echo Momsen's (2010) conclusion that women generally suffer from disasters more than men when they have less access to resources, even though they may play more important roles in coping with and recovering from environmental problems. As an example, during a flood, women's reproductive tasks remain unchanged, but they very often stand last in line when getting assistance if any is sent to the flooded area. Thus, the collective impacts of *food scarcity, being nutritionally disadvantaged, facing livelihood losses,* and other factors can combine to make women suffer much more severely than men after floods (Rashid & Michaud, 2000); this is true in many developing country contexts, even though men are often known to suffer more *loss of life or injuries* during disasters (this is particularly true when they are called upon to attempt heroic acts). Added to this is the impact placed on women, families and communities, as well as to the men directly, when men lose their health and/or livelihoods due to climate change.

Another important aspect worth mentioning in the gender and climate change debate is the decision-making process. Women's voices, and the voices of marginalized men, are often missing in decision-making processes. As a consequence, their interests are likely to be disregarded in policy formulation and development planning.

Taking women's involvement in environmental projects as an example, women are often seen as being potentially key promotional agents for environmental conservation and tree-planting projects without incorporating a careful consideration of their domestic burden. As a result, many have argued that particularly because women are disadvantaged with regard to the gender division of labor (Tirado et al., 2010), adaptation strategies should be designed to lessen and not intensify women's burden in the context of climate change. Moreover, the division of labor between women and men in different sectors of production (e.g., agriculture, collection of NTFPs, hunting and fishing) needs to be analyzed in specific contexts to identify the different degrees of vulnerability women and men face due to the negative effects of climate threats (Denton, 2002). Without a good gender analysis, not only of gender-differentiated livelihoods but also additional responsibilities – listening to both women and men – policymakers might assume that women can play a key role in developing strategies to cope with climate change without taking into account the increased burden those women might have to bear.

Particularly in the context of threats to food insecurity, *livelihood sustainability* – for both women and men – becomes an indispensable strategy in climate-change adaptation. Sustainable livelihoods, at a very basic level, are "about the ways in which individuals prepare for and respond to the effects of environmental change" (Johnson, 2012, p. 6). One such approach is through agricultural intensification, which aims at increasing the value of output per hectare by increasing inputs of labor or capital on a smallholding. This approach is usually considered a positive process that can bring about sustainable livelihoods; however, it is also viewed as a method that is potentially harmful to the sustainability of the environment (Carswell, 1997). Extensification, on the other hand, is a process of increasing productivity by developing a more extensive production system characterized by a larger area of cultivated land and lower input of labor and capital. This approach is considered less applicable in a context of limited available land.

Livelihood diversification "refers to attempts by individuals and households to find new ways to raise incomes and reduce environmental risks, which differ sharply by the degree of freedom of choice (to diversify or not), and the reversibility of the outcome" (Hussein & Nelson, 1998, p. 3). Migration can be used as an example. Migration, including local migration and longer-distance migration, according to Paavola (2008), is a strategy that enables an agricultural household to transform its opportunity set and associated risks. Local migration can be a means to gain access to such assets as land, employment opportunities, markets and other resources. Longer-distance migration can relieve rural credit constraints by generating remittances (Hussein & Nelson, 1998).

In many cases, a single strategy cannot get people through stressful events. In the case of Morogoro, Tanzania (Paavola, 2008), rural populations have applied a series of interventions, including agricultural extensification, agricultural intensification, livelihood diversification, and out-migration to survive through times of crisis. However, they have also suffered environmental consequences that were actually made worse by pursuing some of these strategies. For example, agricultural intensification has been blamed for land degradation, and agricultural extensification will be diminished as land becomes increasingly scarce. Livelihood diversification also may require substantial complementary support to prevent environmental degradation (e.g., of soil, forests and water) as a result of diversification efforts.

For this reason, climate-change adaptation through livelihood diversification can be important, but negative outcomes will have to be strenuously avoided. According to Agrawal et al. (2008), local institutions can play a crucial role in "promoting effective adaptation and enhancing the adaptive capacity of vulnerable rural populations." They emphasize that it is important to enhance local institutional capacities and support a greater role for institutional partnerships in order to achieve effective, and socially and environmentally sound, types of adaptation. CARE International (2011) has also suggested that livelihood strategies should be developed toward diversity because when one of the livelihood strategies is at risk, the household will still have an alternative livelihood to meet food and income needs; however, *sustainability* will also have to be taken as a key goal as households move toward developing these adaptation strategies.

This theme, emphasizing the importance of food and livelihood security, will be central to the analysis of gender and climate change presented here. How men and women contribute – or fail to contribute – to meeting these crucial requirements in the face of climate threats will depend in large part on their designated roles and responsibilities and the degree to which they are open to change. It is to these concerns that we now turn.

Men, masculinities and work in climate change and development discourse

There have been many studies conducted on climate change and development trends. Within this research, women, particularly poor women, have been viewed most often as climate-change victims or victims of development-related concerns, as opposed to agents of change. For this reason, we need to also discuss women's positive contributions to efforts to deal effectively with climate change. Similarly, we also need to ask, what is the effect of climate change on men, and what positive contributions can they make in responding to this process of change?

In many cultures, being a man is synonymous with being the breadwinner or head of the household (Brittan, 2005). Men's sphere of work is normally associated with notions of masculinity, which in many contexts link to an ideology of power and authority (Connell & Messerschmidt, 2005).

Men would normally strive to adapt once they face an economic crisis due to climate threats, disasters or other causes. Along the way, their identities as bread-winners or income earners might be challenged, as would their role in the house-hold. For example, in a case study conducted in Mexico, Gutmann and Viveros (2005) found that it was not unusual to find men who lacked employment alterna-tives and did not have a high level of education or strong economic resources at home caring for their own small children, while in upper-class families, child care was mostly performed by maids and nannies. However, there have been other studies that have argued otherwise; for example, a study conducted in Russia and the United Kingdom showed that when men became unemployed or economically inactive, their contributions to housework and care-related activities changed very little (Ashwin & Lytkina, 2004). Although many factors influence these different responses to change, one involves the conception of masculinity and appropriate masculine behavior in a specific cultural context.

In Viet Nam, particularly in rural areas, men's work is often described as hard work, needing a great deal of physical strength. This work is mostly done outside the home; women's work, in contrast, is often traditionally associated with tasks carried out within or nearby the home. A large part of women's work, both in productive and reproductive spheres, is home-based and is generally either unpaid or poorly paid, which can create a gap between women and men in terms of financial contributions to household income. These differences between women's and men's socially des-ignated responsibilities have created gender-differentiated stereotypical roles and statuses, including specific conceptions of masculinity (Hoang & Yeoh, 2011).

Masculinity is a pattern of behaviors attached to social expectations about how men should act and their position with respect to gender relations. Although "com-mon features" of masculinity generally include "dominance, toughness, and risk-taking" (MenEngage Alliance & UN Women, 2014), masculinity is usually seen as socially constructed so there are "multiple definitions" of masculinity (Connell, 2015). Different cultures, periods of history, or even different groups of men in one society would have very different expressions of masculinity. For example, in some cultures, soldiers are considered heroes, and violence is "the ultimate test of masculinity," while in others, violence is regarded as "contemptible." Moreover, the conception of masculinity in working-class life may be different from the conception of men's positive roles in middle- or upper-class life. Among different masculinities, some are more honored while others can be dishonored. As stated by Kabeer (2007), gender norms and practices related to the division of labor are partly shaped by different cultures, social class, and ethnicity. Men from poor and less-privileged groups would be the least able to conform to their traditional conception of masculinity when their livelihoods are not secure, or their ability to make a living is threatened, with the result that they may not be able to deliver on the breadwinner role they are socially expected to play.

What men do is important in defining the power structure in the family. However, in some cases what men *do not do* is even more important in creating masculine identities (Wetherell & Edley, 1995). Housework, for example, is considered to be within women's sphere in many societies, and men keep their distance from doing domestic work to maintain their status. It is common in many societies for women to be expected to share productive work, but men are not expected to engage on the same level with reproductive work. Men's participation in domestic work is often considered as helping their wives rather than sharing responsibilities. In some contexts, men's participation or non-participation in domestic work and child care identifies men's "degree" of masculinity (Hoang & Yeoh, 2011). In some cases, even when women take over the breadwinner position, men are still reluctant to share housework because of the fear of compromising their sense of masculinity and self-esteem. This was shown in a case study of men in the Philippines – when their masculinity was more secure as a consequence of their financial contribution to the family, they were more willing to participate in domestic work (Parreñas, 2005).

In cases where men were unable to financially support their families, including as a consequence of climate-related events, it was found that the failure to cope with what is perceived as a sudden change or a reduction in their power often led to psychological problems, and thus men in these circumstances can be considered victims as well. In other cases, evidence of domestic violence and drug and alcohol abuse were recorded (Ashwin & Lytkina, 2004; Kabeer, 2007). These were also considered some of contemporary masculinity's "toxic effects," which include negative impacts on the lives of men themselves as well as on women. Physical violence is, according to Connell and Messerschmidt (2005), one expression of toxic practices built on the concept of hegemonic masculinity and "based on practice that permits men's collective dominance over women to continue" (p. 840). Hegemonic masculinity is a common term used widely in gender studies, referring to a culturally dominant form of masculinities not just among different masculinities, but more importantly in the gender order (Connell, 2000).

Engels used male privilege to build the definition of patriarchy in *The Origin of the Family, Private Property and the State* (1884). Back then, patriarchy, which was referred to as "rule by fathers," was a social system in which fathers held primary authority over women and children. Patriarchy implies ideologies that are based on the idea that men rule, and women submit socially, politically and economically. In a patriarchal society, power relations are hierarchical and unequal. In *Theorizing Patriarchy*, Walby defined patriarchy as "a system of social structures and practices in which men dominate, oppress and exploit women" (Walby, 1990, p. 20). Men have power over production, reproduction, sexuality, assets, economic resources and almost all aspects of women's lives. The power relations are maintained by gender stereotypes regarding masculinity and femininity that have been socially and traditionally determined.

In the case of Viet Nam, according to Tran et al. (2006), the mainstream culture is male-centered and strongly patriarchal due to the enduring influence of Confucianism. Men are considered the head of the household with a designation as having the most important role in the family. Men are the ones who are socially designated for practices associated with ancestor worship, and they benefit from patrilocality. In earlier times in Viet Nam, women were clearly constrained by patriarchal perceptions and Confucianism that placed women as men's inferiors and subordinates. The constraints were expressed in the forms of the "three obediences" (subordination to the father and the elder brothers when young, subordination to the husband when married, subordination to the sons when widowed[4]) and the "four moralities" (womanly work, womanly appearance, womanly speech, womanly deeds[5]) – principles that all women should follow. Although modified, the influence of these conceptions continues in modern times. In Viet Nam, patriarchy is used to express the dominance of men over women, especially in rural areas. Patriarchal society nurtures men's power over practically every aspect of women's lives. Gender stereotypes and prejudices are also maintained accordingly; for example, Viet Nam's mainstream society expects women to be beautiful in order to be selected by men because getting married is seen as the biggest task and goal of a woman's life.

Especially after Confucianism was introduced into Viet Nam's culture, the principle of valuing men over women was emphasized (as the expression goes, "One boy child, write 'yes'; ten girl children, write 'no'"). Although it may have originated from a more positive intention, that men should shoulder the family's responsibilities, it led to an overestimation of men's contributions and an underestimation of women's roles in the family. Among other problems, this way of thinking allowed men to have many wives (and "naturally," the reverse was not allowed). Women were often blamed for any problems that happened in their lives; as an example, a woman would be blamed for not being able to keep her husband and her family together if her husband had an affair. A woman could even be blamed for her husband beating her because people might think she was not well behaved or what she did or said upset her husband. Gender stereotypes and gender prejudice about masculinity and femininity have thus established and maintained the "accepted" (traditional mainstream) gender order in Viet Nam with men as rulers and women as subordinates.

Men, however, have not always benefited from such privileges. Men are under pressure to act in "manly" ways that are, in many societies, synonymous with behaviors manifesting physical strength. They are often not encouraged to participate in housework or to be very involved in raising and taking care of their own children but rather to focus on financially supporting their families. They are often raised and taught to be tough and less emotional. If they fail to meet social expectations of masculinity, they would despise themselves as well as be despised by their respective families and communities (FAO & UNDP, 2002; Salemink, 2003; Swinkels & Turk, 2006; Tran et al., 2006).

In the same way, men also have not always benefited from such privileges when viewed from contexts involving climate change and disasters. For example, although much of the literature emphasizes *women's* vulnerability in relation to climate change and disasters related to climate change, recent studies indicate that, in certain contexts, *men's* work-related activities make them more vulnerable to negative health impacts of climate change – for example, through increases in heat exposure and infectious disease – than would be true for women in these settings (Sellers, 2018). Moreover, as mentioned earlier, men often suffer greater injury or loss of life during disasters, depending on what is expected of them in a specific context.

Regarding difficulties, recent research now offers new insights into different responses to disasters using an intersectional approach to understanding men and masculinity (Enarson & Pease, 2016). This includes a focus on the potential for an increase in men's psychological and physical stress, which can result in self-destructive behavior, gender-based violence and other serious consequences in the wake of periodic disasters and long-term climate threats, often because of loss of work and income and their previous roles and status in the face of these sudden changes.

Finally, it is important to note that a number of recent discussions on climate change have also emphasized the need to focus on current and potential new contributions that men and boys can make: "It is important to bring men and boys, as well as women and girls, into the gender-and-environment conversation" (United Nations Environment Program, 2016, p. 209). In fact, a number of new approaches to men's participation in climate change–related efforts have been suggested in recent years (e.g., Hultman & Pulé, 2018). These themes – of difficulties faced by men and boys, as well as positive contributions they can make – will be a key part of the analysis presented in this study of gender and climate change in the context of Viet Nam.

Climate-change adaptation strategies from a gender perspective

Currently, in the field of climate change, along with familiar terms such as "vulnerability," "exposure," and "food security," the terms *adaptive capacity* and/or *resilience* are being widely discussed because they refer to the ability of people affected by climate change to shield themselves and recover from its adverse impacts (Turner et al., 2003). Adaptive capacity is dynamic and has proven to be uneven across and within societies, with many individuals and groups having insufficient capacity to cope with or adapt to climate change. Again, this varies according to economic and natural resources, social networks, entitlements and governance, among other determinants.

Regarding women and climate-change adaptation, UN WomenWatch (2009) summarized in their fact sheet the conclusions that social, economic and political barriers limit women's capacity to effectively cope with and adapt to climate change. As noted earlier, in the case of rural women in developing countries, limited mobility and unequal access to resources and decision-making processes are among the biggest challenges to women's assigned responsibilities to secure water, food and fuel and to complete other tasks. The burden of dealing with climate change–related illnesses in the family (e.g., heat stroke or mosquito- or water-borne diseases) can further limit women's abilities to carry out their designated roles.

Case studies from Bangladesh, Ghana, and Senegal have pointed out similar problems (Dankelman et al., 2008), noting important points directly related to cultural and gender conceptions that women's adaptive capacity is constrained by "social expectations" of "appropriateness" regarding women's actions, particularly with respect to their perceived responsibilities toward the household. This can also be analyzed in terms of cultural and social barriers.

Barker and Buchanan-Barker (2011) found that the ability to deal with risks and the adaptive capacity of every individual and group vary depending on their views, values and beliefs. This results in differential power and access to decision-making processes that can promote a strong adaptive capacity for some people while at the same time constraining it for others. For example, in Bangladesh in 1991, a cyclone and flood caused women's death rates to be almost five times that of men's due to women's limited access to information regarding warnings and also because of their limited mobility as they were not allowed to leave home without a male relative. In addition, they lacked survival skills, such as the ability to swim (Dankelman et al., 2008). Women are also expected to continue carrying out their responsibilities as caregivers in times of disaster and environmental stress in many developing countries. As a result, they are likely to be less mobile, leading to more exposure and limited coping capacity to stresses associated with climate threats (Dankelman & Jansen, 2010; Denton, 2002; Oxfam, 2007, 2008).

For these and other reasons, the burden of coping with and adapting to climate change and serious climate-related events is often found to weigh most heavily on women's shoulders. For example, Adger et al. (2007) argued that in subsistence farming communities in Southern Africa, women were disproportionately burdened with the costs of recovery from drought. A study conducted by Nelson et al. (2002) showed evidence of women having difficulties in coping with the disaster and getting back to work after Hurricane Mitch because of their domestic responsibilities. (In this case, we should note that the women's ongoing problems were compounded by the fact that more men than women are said to have died as a direct consequence of the hurricane, including during rescue efforts; again, suffering and tragedy due to climate change affect both men and women, but often result from different direct causes.)

In terms of coping and adaptive capacity, poverty and lack of resources are again constraining factors that need to be emphasized (Adger et al., 2007). Given the negative impacts of climate change on agricultural production, the need for poor women to work even harder to maintain their families can negatively impact women's health and wellbeing. In addition, it constrains women's available time for public participation as well as involvement in decision-making processes, which may consequently lower women's position in the community in terms of being heard and heeded (Makhabane, 2002). According to Barker and Buchanan-Barker (2011), in a patriarchal society where women rarely have control over agricultural income and a household's financial sources, economic poverty disproportionately affects women. This leads to exclusion from decision-making arenas, legislation and policy frameworks.

Long-term policy and strategic planning can thus play a critical role in achieving sustainable adaptation and avoiding "negative adaptation," which often occurs when the poor have no choice but to alter the ways in which they subsist in order to adapt their livelihoods when they are no longer able to cope with short-term shocks – for example, by taking on precarious, dangerous or degrading types of livelihood activities (Davies & Hossain, 1997). For instance, in a case study of the Bambara of Southern Mali, Toulmin (1992) pointed out that women had fewer opportunities for livelihood diversification than men in responding to threats of droughts, poor and variable rainfall and demographic-related problems, and in some cases, prostitution was resorted to as one type of "income diversification strategy" for women.

According to Davies and Hossain (1997), negative adaptation (or "maladaptation") is likely to be irreversible and frequently does not sustainably reduce vulnerability. Apart from avoiding falling into patterns of negative adaptation, they argue that adaptation strategies should be planned along with a gender analysis that can provide an understanding of (1) gender roles and responsibilities in the household and the community; (2) differences in access to and control over resources and decision-making; and (3) other factors that constrain or facilitate the equal participation of both women and men in community-development processes. It can also help in exploring gender-differentiated capacities, needs, and priorities applied in adaptation strategies.

Here, again, a greater focus on gender roles, relations and hierarchies – and not just on women and men separately or as homogeneous entities – will be needed to understand both existing patterns and new possibilities for climate-change adaptation strategies. As MenEngage Alliance (n.d.) notes: "At the same time, an understanding of boys' and men's multiple roles in climate change have remained almost invisible, except in certain areas of research. . . . Boys and men must be part of the solution to achieve gender-informed climate justice. . . . There is little recognition that men's diversity – according to social class, ethnic group, sexuality and other factors – also affects not only the way they live their lives, but the way that they drive or respond to climate change" (p. 2). By recognizing diversity and different experiences and skillsets in this way, the complementary knowledge and abilities of both women and men can be engaged in the crucial search for effective responses to climate threats and positive adaptation strategies that work for the benefit of the entire population, as will be discussed later in the context of ethnic minorities in Viet Nam. It is to a consideration of indigenous groups and ethnic minorities that we now turn.

Indigenous communities and ethnic minorities: analyses of vulnerabilities

According to the Indigenous and Tribal Peoples Convention of 1989 (Geneva, 76th ILC Session, June 27, 1989), "tribal peoples are those whose social, cultural and economic conditions distinguish them from other sections of the national community, and whose status is regulated wholly or partially by their own customs or traditions or by special laws or regulations." Another definition was given by the United Nations during the preparation for the Permanent Forum on Indigenous Issues (New York, January 19–21, 2004); here, indigenous communities, peoples, and nations were defined as "[t]hose which, having a historical continuity with pre-invasion and pre-colonial societies that developed on their territories, consider themselves distinct from other sectors of the societies now prevailing in those territories or parts of them. They form non-dominant sectors of society and are determined to preserve, develop and transmit to future generations their ancestral territories, and their ethnic identity, as the basis of their continued existence as peoples in accordance with their own cultural patterns, social institutions and legal systems" (UNDP, 2004, p. 2). In earlier definitions, indigenous people were often assumed to rely primarily on a subsistence economy; in recent years, however, many indigenous people have shifted to market economies and are exposed to often conflicting outside influences (van Leeuwen, 1998). Nevertheless, definitions of indigenous communities usually include the following components: (1) indigenous people have a distinct culture and ethnic identity; (2) they are considered autochthonous (i.e., not descended from [relatively recent] migrants or colonists); and (3) they possess distinctive social, cultural, political and economic institutions and legal systems.

Viet Nam is a multi-ethnic country, with 54 officially recognized ethnic groups. The largest ethnic group is the Kinh (Viet), making up approximately 85.5% of the population. Fifty-three other ethnic minority groups account for 14.5% of the population, which includes 6.72 million men (50.21%) and 6.66 million women (49.79%). The Tay (1.76 million), Thai (1.72 million), Muong (1.39 million), Khmer (1.29 million) and Hmong (1.25 million) are some of the largest ethnic minorities (UN Women, 2017). With the exception of the Kinh-Hoa (the latter indicating Chinese descent), Khmer, Cham and Muong, the remaining 50 ethnic groups are living in relatively remote mountainous and rural areas, and in many ways are economically and socially marginalized (Baulch et al., 2008). Central Viet Nam, where Nam Giang District is located, consists of many indigenous mountain communities, including the Co Tu ethnic minority group.

According to Plant (2002), there is no clear understanding of the concept and coverage of ethnic minorities (also referred to as "indigenous communities" by many analysts) in Viet Nam. The National Program of Ethnic Classification was set up by Vietnamese ethnologists of the Institute of Ethnology to classify and define ethnic groups (*dân tộc*). Their definition of an ethnic group was as follows:

> *A stable or relatively stable group of people formed over a historical period, based on common territorial ties, economic activities, and cultural characteristics; on the basis of these common characteristics arises an awareness of ethnic identity and a name of one's own.*
>
> *(Plant, 2002, p. 21)*

However, the present classification system has many weaknesses since many groups and sub-groups are not included. Some groups are also classified under the same ethnic label, even while there are greater cultural differences within one group than between two separate ethnic groups. Furthermore, because of cultural, historical and linguistic differences, many smaller ethnic sub-groups are dissatisfied when they are placed within a larger ethnic group rather than being recognized as separate groups (Plant, 2002). Finally, classifications of ethnic minority boundaries sometimes become unclear because, throughout the years, many Kinh people have migrated to areas where many ethnic minorities live, and territorial ties, economic activities and cultural characteristics have become more difficult to distinguish. For example, in the Central Highlands, the Kinh people accounted for 5% of the total population in 1945, 50% in 1975 and more than 70% in 2002 (Plant, 2002).

It is thus difficult to generalize regarding indigenous mountainous communities – that is, the ethnic minority groups that live in relatively remote mountainous regions – in Viet Nam. Salemink (2003) tried to identify some of the ideas most commonly held about these ethnic minorities; however, it is important to keep in mind that ethnic minorities in Viet Nam include culturally, religiously and linguistically diverse populations. First of all, a common notion in Viet Nam is that most ethnic minorities are nomadic (or at least semi-nomadic), who practice swidden (shifting, or slash and burn) cultivation and move on when the soil is "exhausted." Another common assumption held by outsiders regarding ethnic minorities has been that they consist of clearly distinct "tribes" distinguishable by their languages, traditional clothes, architecture and their "manners and customs." However, the reality is much more complex. For example, there have been supra-village institutions allowing different villages to share a common territory for swidden cultivation, even when their local political systems were distinct from one another, reducing the separate features of each group (Epprecht et al., 2011; Imai et al., 2007; Plant, 2002; Swinkels & Turk, 2006).

In addition, many ethnic minority groups were blamed for heavy deforestation in Viet Nam. For this reason, the ethnic minorities have been, since colonial times, encouraged and even forced to move from swidden cultivation to more "modern" and sedentary cultivation. However, these "modern" methods usually did not take sufficient account of the ecology of tropical mountain areas. According to observers, ethnic minorities in Viet Nam typically cleared plots with a controlled use of fire for multiple crops; after two or three years, they cleared a new plot and let the previously cultivated land lie fallow to restore its fertility. The fallow land was then used for grazing cattle and cultivating low-intensity crops, such as grasses or "green manure" (weeds that quickly regenerate soil fertility). After 10–20 years, the cleared plot was regenerated and the same process started again, with the cultivation of multiple crops followed by a period of rest. In this way, the ethnic minority groups made use of a long fallow and rotating farming system (Jones et al., 2002). Most communities that were engaged in swidden cultivation had "a delineated territory and an elaborate local knowledge and regulatory institutions [defined as customary laws] by which suitable, sufficiently regenerated plots of land were periodically reallocated to households to clear and work" (Salemink, 2003, p. 32). Agriculture was and is usually combined with animal husbandry, hunting, fishing and gathering of timber and NTFPs in these cases (Epprecht et al., 2011; Imai et al., 2007; Jones et al., 2002; Plant, 2002).

Regarding spiritual practices, ethnic minorities in Viet Nam usually have been able to conserve their traditional religions in a region where a number of religions (Buddhism, Christianity and others) predominate. Their traditional beliefs and practices are often labeled as "animist" and have been considered by many to be invalid and not "modern" (Salemink, 2003). They generally believe in a number of gods, such as the God of Thunder, and spirits connected with particular places (sacred forests, rivers, mountains, rocks and other natural phenomena) or particular animals (such as pythons). Both women and men with special skills and knowledge become shamans or spirit mediums. Religious beliefs are closely connected with customary laws governing the relations between people, the environment, and the spirit world (Writenet, 2002). Therefore, within their territory, ethnic minority communities have had specific laws on making use of the environment in order to maintain a sustainable equilibrium (Vuong, 2008).

Many studies in different parts of the world have shown that ethnic minorities concentrated in relatively remote upland and mountainous areas are considered to be much poorer than the ethnic majorities because they lack access to infrastructure, health services and educational facilities. This is also true for Viet Nam, where 30% of the country's poor were ethnic minorities even though they only accounted for approximately 14% of the total population, according to a national survey in 2002. The poverty headcount ratio in 2002 was 64.3% for the ethnic minorities and 22.3% for the ethnic majority Kinh-Hoa population (UN Women, 2017). The national survey in 2015 showed that the average income of ethnic minority households in 2015 was 1.161 million Vietnamese Dong (VND) per month (52 USD, using exchange rates at the time), which was equal to 45% of the national average overall (2.605 million VND), and equivalent to 41% (2.888 million VND) of the average income of the Kinh ethnic majority (UN Women, 2017). Although Viet Nam has experienced a remarkable reduction in poverty, it has been argued that the ethnic majorities were the ones who mainly benefited from it. For example, the share of ethnic minorities among the poor accounted for only 20% in 1990, but went up to 30% in 2002 (Imai et al., 2007).

It should be noted that, from the viewpoint of some observers (including many anthropologists), indigenous cultures should not be viewed as "poor" – it is argued that they have resources and assets but not in the sense of a more commercially-oriented type of society. However, these observers would agree that most indigenous communities are at a great disadvantage in terms of control when they move away from relative self-sufficiency and self-determination and are required to move toward full engagement with the mainstream economy and society and become subject to its laws and influenced by its values.

With this in mind, the Government of Viet Nam has implemented many projects and programs aimed specifically at reducing poverty among ethnic minorities. However, many have involved "one size fits all" models that did not have the intended results for all such groups, particularly those facing multiple challenges – not the least of which are climate threats in these remote mountain areas. Many analysts think that new approaches of a more diverse and nuanced nature will help address the specific needs and circumstances of ethnic minority populations in different natural and social environments within the country, and again, this will be taken up in detail in the current study.

Indigenous groups, gender and climate change

Climate change is by no means a new phenomenon. Humankind has always faced and adapted to climate-related changes. Indigenous populations can offer important lessons regarding adaptation strategies and long-term observations of climate change because, in many cases, they have knowledge about environmental trends that has been passed down over a long period of time. Furthermore, indigenous and local communities are often strongly affected by and are highly vulnerable to climate-change impacts.

However, there has been relatively little attention paid to the impact of climate change on indigenous communities as well as their ways of adapting to climate change. This was significantly improved in the "Fourth Assessment Report (AR4) Climate Change 2007: Impacts, Adaptation, and Vulnerability" of the United Nations' Intergovernmental Panel on Climate Change (Parry et al., 2007). The report emphasized that disadvantages faced by indigenous peoples should be taken into consideration in climate-change research and policymaking processes. The authors also mentioned the importance of studying indigenous knowledge systems, which can be valuable sources of information on climate-change impacts and adaptation strategies. A call of the International Alliance of Indigenous and Tribal Peoples of the Tropical Forests was announced in the 11th Meeting of the Conference of the Parties to the United Nations Framework Convention on Climate Change (COP, associated with the UNFCCC, held in Canada in 2005), with the main purpose of supporting and enhancing their full and effective participation at all levels of discussion, decision-making, and implementation. Many analysts concerned with the impact of climate change argue that capacity building for indigenous communities is an area that needs to be funded and paid adequate attention by the international community (IAITPTF/International Alliance of Indigenous and Tribal Peoples of the Tropical Forests, 2005).

As damage caused by climate-change impacts is uneven across different populations, Macchi et al. (2008) put together a list of vulnerability factors that exacerbate indigenous communities' vulnerability in the context of climate change. Indigenous and local communities are directly affected by two types of factors: (1) social vulnerability factors (poverty, inequality, marginalization, nutrition, land tenure and access rights, among others), and (2) biophysical factors (housing quality, land use, land cover change and availability of natural resources). Case studies of indigenous peoples in the Arctic, in Bangladesh, in the tropical rainforest belt, in arid lands and in watersheds were presented to prove that under impacts of climate change, indigenous communities all over the world have been affected and have had to deal with the negative consequences in different ways (Macchi et al., 2008).

Green et al. (2009) conducted a study of indigenous communities in the northern part of Australia. The study affirmed that their isolated location partly caused frequent difficulties in accessing basic housing, infrastructure and community services, which increased their vulnerability to climate-change impacts. In addition, these communities relied on the non-market or so-called customary sector, which was highly dependent on natural resources and therefore was profoundly affected by climate change.

The authors recommended that indigenous communities should be strengthened toward becoming more self-sufficient and resilient; in other words, they opposed strategies that made the communities less self-sufficient and resilient through attempts to further integrate them into the mainstream economy and society in spite of their difficulties in making this transition. Among other concerns, the authors argued that rapid integration into mainstream society can precipitate social dislocation through changes in basic values, practices and relationships. It can erode local systems of knowledge as well as devalue fundamental conceptions of the natural and spiritual world, and it can result in a breakdown of social cohesion and increase dysfunctional behavior, particularly if the pace of change is very rapid. For this reason, the authors argued that understanding indigenous communities' experiences related to climate-change impacts and adaptation, along with a better understanding of new vulnerabilities and new possible responses, is an indispensable source for building greater resilience and social cohesion, given the increase in intensity of climate threats facing indigenous communities at the present time.

In addition to understanding climate-related threats at the community level, the challenges faced by indigenous women and men in responding to climate threats have also been taken up by in recent years using a gender perspective (e.g., McLeod et al., 2018; Sujakhu et al., 2019). In addition, Vinyeta et al. (2015) analyzed the impacts of climate change on indigenous communities in the United States, particularly the negative impacts on not only women and girls but also men and boys through challenges to their roles, livelihoods and indigenous knowledge.

Future studies that are more focused on the interactions between gender, age group, income and other sources of differentiation will aid in bringing out vulnerabilities, worldviews, aspirations and circumstances that distinguish different segments of indigenous and ethnic minority communities. These nuanced studies will then be able to help guide responses in more effective ways.

As mentioned earlier, it is also important to note the large body of literature concerned with indigenous communities and climate change that focuses on local/indigenous knowledge in particular. Some studies emphasize the central importance of indigenous knowledge in allowing more effective responses to climate change, drawing on the knowledge of those (both men and women) who have the deepest understanding of their natural environments (forest transitions, seed varieties, animal behavior, the use of intercropping and other forms of knowledge and techniques). In other contexts, however, rapid climate change is said to have undercut the indigenous knowledge passed down through generations: as David Pulkol from the Karamajong community in Uganda noted, "In our community the elders interpret certain signs from nature to know when to plant their crops or when to start the hunting season. But with climate change it is becoming impossible to make such predictions" (Minority Rights Group International, 2008, p. 1).

Gender is important in analyses of indigenous knowledge as well. In particular, recognizing women's and men's indigenous/local knowledge is important in order to draw on all sources of relevant perspectives. Among other types of knowledge, women are likely to have a greater understanding of certain parts of the local natural environment due to their livelihood activities (e.g., certain types of botanical knowledge in fields and forests), but in a rigidly patriarchal system, their knowledge is often neglected. As an example, in an island in the Federal States of Micronesia, women were the ones who used their knowledge of island hydrology to find potable water by digging wells that reached freshwater sources during a drought (Institute of Development Studies, 2011); however, in a context with a more pronounced gender hierarchy, the women's knowledge could easily have been dismissed as irrelevant or uninformed.

Continuing studies of local knowledge held by different groups and individuals within indigenous and ethnic minority communities will certainly be an important part of an effort toward strengthening community cohesion and resilience, both in terms of practical responses to climate change and, in some cases, maintaining insights into natural processes that might not be noticed or appreciated when approaching the natural environment from a strictly industrial or commercial point of view. This approach to understanding the natural world stands in contrast to the assumption that local knowledge is either not valuable or is no longer applicable under present circumstances. In fact, as will be discussed in the following chapters, a blending of "traditional" and "modern" sources of knowledge is sometimes the most useful approach of all, depending on specific conditions and goals.

Gender and climate change in the context of Viet Nam

Being aware of the existing and potential impacts of climate change, Viet Nam's central government has taken problems of climate-change adaptation explicitly into its planning processes in recent years. It was resolutely stated in the 2005 Law on Environmental Protection that the government was willing to formulate commitments and perform its international responsibilities related to climate change. Viet Nam signed the United Nations Framework Convention on Climate Change (UNFCCC) treaty in 1992 and the Kyoto Protocol in 1998. These efforts were demonstrated by many policies and legal documents that have been issued in recent years, such as Decision no. 35/2005/CT-TTg, dated October 17, 2005, of the Prime Minister on implementing the Kyoto Protocol signed in 2002 under the UNFCCC; Decision no. 47/2007/QĐ-TTg, dated April 6, 2007, of the Prime Minister assigning responsibility to the Ministry of Natural Resources and Environment (MONRE) and other ministries and local departments regarding implementation of the Kyoto Protocol and Clean Development Mechanism; and Decision no. 60/2007/NQ-CP, dated December 3, 2007, of the government entrusting MONRE, in coordination with relevant ministries and departments, with developing a National Program on Climate Change Adaptation. The country is also implementing its Nationally Determined Contribution (NDC) with the target of reducing greenhouse gas (GHG) emissions by 8% by 2030 (i.e., as an unconditional contribution), or by 25% with international support (as a conditional contribution).

From a gender perspective, the 2015 Law on Promulgation of Legal Documents no. 80/2015/QH13 (Article 5, Item 4) and the Law on Gender Equality no. 73/2006/QH11 (Article 21) stated that gender equality mainstreaming needs to be institutionalized in the process of making and approving a legal document. The gender-equality goals are also stipulated in the 2015 Law on Promulgation of Legal Documents as one of the governing guidelines of national budget planning and expenditure (Article 8, Item 5). In the context of Viet Nam aspiring to achieve the Sustainable Development Goals (SDGs), as stated in the National Action Plan for the Sustainable Development Agenda 2030, gender equality is a human right and cannot be separated from the challenges of poverty, hunger and environmental issues. However, since gender mainstreaming has come relatively recently to the country, both the collection of data and integration into policy and practice are incomplete. Moreover, gender stereotypes persist and negatively impact women's advancement, especially within disaster risk reduction and resilience building, and there continues to be unequal access to control over land and productive assets, training, information, technology, extension services and finance.

The NDC plan is an example of this as-yet-limited approach to gender concerns. Viet Nam signed the Sendai Framework in 2015 and ratified the Paris Agreement on November 3, 2016, wherein 196 parties agreed to transform their development trajectories so that the world can achieve sustainable development by limiting warming to 1.5–2°C above pre-industrial levels. This document detailed the government's commitment and strategic plan to reduce GHG emissions and implement the adaptation plan under the UNFCCC. Although Viet Nam's 2011 National Climate Change Strategy identified gender equality as one of its priority areas, gender was not thoroughly integrated into the document, but instead appeared only in limited sections related to adaptation.

Despite the large amount of research confirming the serious gender implications of climate-change impacts, we find that legal documents, policies, strategies and programs aiming to address climate-change issues and gender dimensions of climate change in Viet Nam remain incomplete. It is often stated in the documentation on the gender aspects of rural livelihoods that climate-change impacts as well as climate-change responses are not gender neutral. In spite of this, there is a notable lack of adequate consideration of gender relations and especially women's interests, specifically in climate change–related policies and strategies adapting to climate threats.

Moreover, given the course of economic development and increasing women's roles in productive as well as reproductive spheres over time, there has been a growing desire among both citizens and policymakers in many different country contexts to challenge the rationale of patriarchal systems and culture, and more specifically to challenge men's power and privilege over women (MenEngage Alliance & UN Women, 2014). The combination of increased women's participation in the labor force and the call for gender equality have challenged the ideology of patriarchy; in Viet Nam as well, since Doi Moi reforms were instituted in the mid-1980s, women have increasingly gained more power and freedom, and in the process they have also proven their important contributions to society. As a result, many argue that the patriarchal system has now, more than ever, been seriously challenged in national deliberations in Viet Nam and faces a great number of changes related to labor and power relationships (FAO & UNDP, 2002; Tran et al., 2006).

Because of the gender implications of climate change and climate-change responses, gender-sensitive planning and policymaking in Viet Nam are undeniably needed at all of the relevant sectoral, regional and community levels (Oxfam & UN-Vietnam, 2009). Moreover, as noted earlier, differences across women (and also across men) based on ethnicity, age, household headship, income status and other factors will need to be taken into consideration as new policies and programs are devised (Huynh & Resurreccion, 2014). With this in mind, some of the current policies and new policy proposals being advanced in Viet Nam will be considered in detail in the chapters that follow, particularly in the concluding chapter (Chapter 5).

Research methodologies used in this study

Building on the previous discussions, this study has aimed to understand (1) the impacts of changes in climate, regulation and society that a Co Tu ethnic minority community – including both women and men of different age groups – has experienced over the past two to three decades; (2) how Co Tu women and men have responded to these changes; and (3) how the responses of Co Tu women and men have, in turn, challenged existing gender roles, relations and hierarchies and with what implications for present and future generations of the Co Tu community (middle-aged women and men, young women and men, and the elderly men and women in the community). We will now turn to a discussion of the research methodologies that were used to address these questions, along with the reasons for selecting this study area – including an introduction to the serious problems faced by the community in the context of a rapidly changing environment and society.

Methodologies and research strategies

The present study has benefited from the use of both qualitative and quantitative research methods and data collected over an extended period of rapid change (from 2011 to 2019). It has employed four data-collection methods: key informant interviews (KIIs), focus group discussions (FGDs), in-depth interviews (IDIs), and a household survey. With the main purpose of examining changes over time, data were collected three times: (1) the first and primary period of data collection was undertaken during visits in 2011 and 2012, making use of all four collection methods, with a focus on middle-aged and older women and men who experienced changes in climate and climate-related events; (2) the second period of data collection was conducted in 2014 with a focus on IDIs with younger women and men in order to explore impacts of socioeconomic changes on the next generations' lives, prospects and livelihoods; and (3) the most recent visits for data collection were conducted in 2019, focusing on IDIs with young, middle-aged and older women and men and FGDs with young and middle-aged women and men. These last visits served as a way to verify the earlier findings and update the data obtained in the first two periods of data collection in the community.

Data-collection efforts in 2011 and 2012 included 22 KIIs with government officials and representatives of eight villages in Ca Dy Commune; six FGDs undertaken with two groups of six female respondents, two groups of six male respondents and two mixed groups with four male and four female respondents; 64 IDIs with 32 female and 32 male villagers in eight villages; and a household survey conducted in 2012 that covered 300 households from eight villages. The 2014 data-collection procedure was simpler, with eight IDIs with young women and young men. In 2019, three FGDs were conducted with young women, young men, and middle-aged women; in addition, five IDIs were conducted with young women, seven IDIs were conducted with young men, four IDIs were conducted with middle-aged women, and three IDIs were conducted with elders in the community, in addition to later follow-up questions asked by local research team members.

Given the long distance from one village to another and the limitations of many of the residents' ability to understand and speak the Kinh (national Vietnamese) language, the hesitation of the local Co Tu community in communicating with outsiders and the reliance of the research team on the support of local authorities and village patriarchs, convenience sampling was applied in selecting respondents for the household survey, IDIs, and FGDs. Respondents for KIIs were selected using a convenience-sampling technique that allowed the researchers to reach respondents in circumstances in which available respondents were few in number. Interviews also required a high degree of trust to initiate contact (Atkinson & Flint, 2001), which was facilitated by using local contacts and local research assistants as part of the research team.

Data analysis and structure of the book

This study applied different approaches and techniques for qualitative and quantitative data collection and analysis. However, given the inherent nature of a social and cultural study concerned with gender relations, a stronger focus was put on collecting, analyzing and presenting the qualitative data. By using a data-deduction technique, which is "a form of analysis that sharpens, sorts, focuses, discards, and organizes data in such a way that 'final' conclusions can be drawn and verified" (Namey et al., 2008), the research analyzed qualitative data for content and thematic analyses in order to produce meaningful explanations regarding the nature of the analyzed phenomenon or evidence. Qualitative data were supplemented and in part verified by quantitative information.

As one of the primary objectives of the research has been to investigate the dynamics and effects of the Co Tu women's and men's adaptation strategies to their rapidly changing climate and society, both descriptive and explanatory research methodologies were applied to this study. First, the study attempted to describe the impacts of climate and regulatory changes on the Co Tu community's food and livelihood security, given changes in society and development pressures. Second, the strategies that allowed them to cope with changes were analyzed from a gender perspective by examining the changing gender division of labor and gender-differentiated roles throughout these years. Third, gender-specific outcomes of adaptation strategies were analyzed with the aim of investigating changing women's and men's roles and voices in the family and in the community, and changes in the prevailing gender hierarchies.

In addition to using people's perceptions examined through qualitative and quantitative primary data, the research also used secondary data to cover two key areas: (1) the socioeconomic context of the studied area as a foundation for research, and (2) climate-related data in order to define whether or not climate has actually changed based on scientific evidence, thereby not relying solely on people's perceptions that might be influenced by their own personal feelings and experiences. Meteorological data were collected from the Tra My and Tam Ky meteorological stations in Quang Nam Province, which were available (depending on the data) covering the period from 1980 to 2005 or 2008. The meteorological data, particularly on rainfall and temperature patterns, were also used to identify variation trends by using a Mann-Kendall test (MK test). This method is simple and robust and handles missing values as well as values below a detection limit (Kahya & Kalayci, 2003). This initial analysis was then supplemented by more recent data, as well as projections regarding climate trends over the coming decades.

The following chapter (Chapter 2) will begin with a further introduction to the Co Tu community of Ca Dy Commune, including the traditional gender hierarchy and patriarchal system. It will then present an analysis of climate change–related meteorological data from the research area in recent decades, comparing the data with local perceptions – and here we will find that there are often differences in perceptions according to gender, given the different roles played by women and men in the community. Chapters 3 and 4 will then focus on regulatory policy and socioeconomic changes that compound the serious consequences of climate change and have rendered unviable the earlier gender-based livelihood patterns, thus requiring major changes in gender roles and, in so doing, challenging existing gender hierarchies.

Chapter 5 will conclude with a summary of main findings and how these findings tie in with the literature discussed in this and the following chapters. Moving beyond this case study, some initial implications will be discussed regarding the serious problems that often arise in relatively isolated communities that are suddenly forced into the mainstream economy and society in the wake of climate threats and other compounding factors. This discussion will take into account a number of different perspectives on how to deal with these problems, as well as new opportunities created by the combined forces of climate, regulatory and social change that, in such a short period of time, have disrupted and now are likely to further challenge earlier gender-based roles, relations and hierarchies in these rapidly changing and vulnerable communities.

Notes

1 An article summarizing one part of the early thesis findings can be found in Pham et al. (2016).
2 The Kinh, comprising more than 85% of the national population, is the main ethnic majority group in Viet Nam. In Viet Nam, the Hoa (Chinese descent) ethnic group, although small, is often grouped with the Kinh as part of the Kinh-Hoa ethnic majority.
3 Members of the Co Tu ethnic minority group – also known as the Katu, Ca Tu, Gao, Ha, Phuong, and Ca-tang – live primarily in Central Viet Nam and Lao PDR. The Co Tu ethnic minority is one of 54 officially recognized ethnic groups in Viet Nam.
4 Tại gia tòng phụ. Xuất giá tòng phu. Phu tử tòng tử.
5 Công, Dung, Ngôn, Hạnh

References

Abeygunawardena, P., Vyas, Y., Knill, P., Foy, T., Harrold, M., Steele, P., Tanner, T., Hirsch, D., Oosterman, M., Rooimans, J., Debois, M., Lamin, M., Liptow, H., Mausolf, E., Verheyen, R., Agrawala, S., Caspary, G., Paris, R., Kashyap, A., Sharma, A., Mathur, A., Sharma, M., & Sperling, F. (2009). *Poverty and climate change: Reducing the vulnerability of the poor through adaptation*. World Bank. http://documents.worldbank.org/curated/en/534871468155709473/Poverty-and-climate-change-reducing-the-vulnerability-of-the-poor-through-adaptation

Adger, W. N. (1999). Social vulnerability to climate change and extremes in coastal Vietnam. *World Development, 27*(20), 249–269. https://doi.org/10.1016/S0305-750X(98)00136-3

Adger, W. N., Agrawala, S., Mirza, M. M. Q., Conde, C., O'Brien, K., Pulhin, J., Pulwarty, R., Smit, B., & Takahashi, K. (2007). Assessment of adaptation practices, options, constraints and capacity. In M. L. Parry, O. F. Canziani, J. P. Palutikof, P. J. van der Linden, & C. E. Hanson (Eds.), *Climate change 2007: Impacts, adaptation and vulnerability. Contribution of working group II to the fourth assessment report of the intergovernmental panel on climate change* (pp. 717–743). Cambridge: Cambridge University Press. www.ipcc.ch/site/assets/uploads/2018/02/ar4-wg2-chapter17-1.pdf

Agrawal, A., Kononen, M., & Perrin, N. (2008). *The role of local institutions in adaptation to climate change*. Paper presented at the Social Dimensions of Climate Change, World Bank. https://doi.org/10.1596/28274

Alston, M. (2015). *Women and climate change in Bangladesh*. London: Routledge. https://doi.org/10.4324/9781315774589

Alston, M., & Whittenbury, K. (Eds.). (2013). *Research, action and policy: Addressing the gendered impacts of climate change*. Dordrecht: Springer. https://doi.org/10.1007/978-94-007-5518-5

Altieri, M., & Koohafkan, P. (2008). *Enduring farms: Climate change, smallholders and traditional farming communities*. Third World Network. www.agroeco.org/doc/Enduring%20farms.pdf

Arora-Jonsson, S. (2011). Virtue and vulnerability: Discourses on women, gender and climate change. *Global Environmental Change, 21*(2), 744–751. https://doi.org/10.1016/j.gloenvcha.2011.01.005

Ashwin, S., & Lytkina, T. (2004). Men in crisis in Russia: The role of domestic marginalization. *Gender & Society*, *18*(2), 189–206. https://doi.org/10.1177/0891243203261263

Atkinson, R., & Flint, J. (2001). Accessing hidden and hard-to-reach populations: Snowball research strategies. In *Social research update* (vol. 33). Guildford: University of Surrey.

Barker, P., & Buchanan-Barker, P. (2011). Myth of mental health nursing and the challenge of recovery. *International Journal of Mental Health Nursing*, *20*(5), 337–344. https://doi.org/10.1111/j.1447-0349.2010.00734.x

Baulch, B., Hansen, H., Trung, L. D., & Tam, T. N. M. (2008). The spatial integration of paddy markets in Vietnam. *Journal of Agricultural Economics*, *59*(2), 271–295. https://doi.org/10.1111/j.1477-9552.2007.00148.x

Brittan, A. (2005). *Masculinity and power*. Oxford: Basil Blackwell.

Brody, A., Demetriades, J., & Esplen, E. (2008). *Gender and climate change: Mapping the linkages – a scoping study on knowledge and gaps*. Brighton: Bridge, Institute of Development Studies.

Buckingham, S., & Le Masson, V. (2017). *Understanding climate change through gender relations*. London: Routledge. https://doi.org/10.4324/9781315661605

Cannon, T. (2002). Gender and climate hazards in Bangladesh. *Gender & Development*, *10*(2), 45–50. Oxford: Oxfam. https://doi.org/10.1080/13552070215906

CARE International. (2011). *Annual report and accounts for the year ended 30th June 2011*. www.careinternational.org.uk/sites/default/files/CIUK-annual-report-and-accounts-2011.pdf

Carr, E. R., & Thompson, M. C. (2014). Gender and climate change adaptation change in agrarian settings: Current settings, new directions and research frontiers. *Geography Compass*, *8*(3), 182–197. https://doi.org/10.1111/gec3.12121

Carswell, G. (1997). *Agricultural intensification and rural sustainable livelihoods: A "think piece"*. Institute of Development Studies Working Paper No. 64. IDS. www.ids.ac.uk/files/dmfile/Wp64.pdf

Chambers, R., & Conway, G. R. (1992). *Sustainable rural livelihoods: Practical concepts for the 21st century*. Institute of Development Studies Discussion Paper No. 296. IDS. www.ids.ac.uk/files/Dp296.pdf

Chaudhry, P., & Ruysschaert, G. (2007). *Climate change and human development in Vietnam*. Human Development Occasional Papers (1992–2007), HDOCP-2007-46. Human Development Report Office, United Nations Development Programme. https://EconPapers.repec.org/RePEc:hdr:hdocpa:hdocpa-2007-46

Colfer, C. J. P., Sijapati Basnett, B., & Elias, M. (Eds.). (2016). *Gender and forests: Climate change, tenure, value chains and emerging issues*. London: Routledge.

Connell, R. W. (2000). *The men and the boys*. Berkeley: University of California Press.

Connell, R. W. (2015). Masculinities: The field of knowledge. In S. Horlacher (Ed.), *Configuring masculinity in theory and literary practice*. London: Brill Rodopi. https://doi.org/10.1163/9789004299009

Connell, R. W., & Messerschmidt, J. W. (2005). Hegemonic masculinity: Rethinking the concept. *Gender and Society*, *19*(6), 829–859. https://doi.org/10.1177/0891243205278639

Dankelman, I. (2010). Introduction: Exploring gender, environment and climate change. In I. Dankelman (Ed.), *Gender and climate change: An introduction* (pp. 1–20). London: Routledge. https://doi.org/10.4324/9781849775274

Dankelman, I., Alam, K., Ahmed, W. B., Gueye, Y. D., Fatema, N., & Mensah-Kutin, R. (2008). *Gender, climate change and human security lessons from Bangladesh, Ghana and Senegal*. The Women's Environment and Development Organization (WEDO) with

ABANTU for ELIAMEP for Development in Ghana, ActionAid Bangladesh and ENDA in Senegal. www.gdnonline.org/resources/WEDO_Gender_CC_Human_Security.pdf

Dankelman, I., & Jansen, W. (2010). Gender, environment, and climate change: Understanding the linkages. In I. Dankelman (Ed.), *Gender and climate change: An introduction.* (pp. 21–54). London: Routledge. https://doi.org/10.4324/9781849775274

Dasgupta, S., Laplante, B., Meisner, C., Wheeler, D., & Jianping, Y. (2007). *The impact of sea level rise on new countries: A comparative analysis.* World Bank Policy Research Working Paper 4136. World Bank. http://documents.worldbank.org/curated/en/156401468136816684/pdf/wps4136.pdf

Davies, S., & Hossain, N. (1997). *Livelihood adaptation, public action and civil society: A review of the literature.* Institute of Development Studies Working Paper 57. www.ids.ac.uk/files/Wp57.pdf

Denton, F. (2000). Gender impact of climate change: A human security dimension. *Energia News, 3*(3), 13–14.

Denton, F. (2002). Climate change vulnerability, impacts, and adaptation: Why does gender matter? *Gender and Development, 10*(2), 10–20. https://doi.org/10.1080/13552070215903

Eckstein, D., Hutfils, M., & Winges, M. (2018). *Global climate risk 2019: Who suffers most from extreme weather events? Weather-related loss events in 2017 and 1998 to 2017* Briefing Paper. Global Climate Risk Index 2019. Bonn: Germanwatch.

Enarson, E., & Pease, B. (Eds.). (2016). *Men, masculinities and disaster.* London: Routledge.

Engels, F. (1884). The origin of the family, private property, and the state. In *Marx/Engels selected works* (vol. 3). www.marxists.org/archive/marx/works/download/pdf/origin_family.pdf

Epprecht, M., Müller, D., & Minot, N. (2011). How remote are Vietnam's ethnic minorities? An analysis of spatial patterns of poverty and inequality. *The Annals of Regional Science, 46*(2), 349–368. Springer-Verlag (Original work published in 2009). https://doi.org/10.1007/s00168-009-0330-7

FAO. (2016). *"El Niño" event in Vietnam agriculture, food security and livelihood needs assessment in response to drought and salt water intrusion.* www.fao.org/3/a-i6020e.pdf

FAO. (2017). *The impact of disasters and crises on agriculture and food security.* (Figures 2 and 3). http://www.fao.org/3/I8656EN/i8656en.pdf

FAO, IFAD, UNICEF, WFP, & WHO. (2018). *The state of food security and nutrition in the world 2018: Building climate resilience for food security and nutrition.* www.who.int/nutrition/publications/foodsecurity/state-food-security-nutrition-2018/en/

FAO, & UNDP. (2002). Gender differences in the transitional economy of Vietnam. In *Key gender findings: Second Vietnam living standards survey, 1997–98.* FAO. www.fao.org/3/AC685E/ac685e00.htm

Green, D., Jackson, S., & Morrison, J. (2009). *Risks from climate change to indigenous communities in the tropical North of Australia.* Canberra: Department of Climate Change and Energy Efficiency.

Gutmann, M., & Viveros, M. (2005). Masculinities in Latin America. In M. Kimmel, R. W. Connell, & J. Hearns (Eds.), *Handbook of studies on men and masculinities* (pp. 114–128). Thousand Oaks, CA: Sage Publications. http://doi.org/10.2139/ssrn.2293658

Hoang, L. A., & Yeoh, B. S. A. (2011). Breadwinning wives and "left behind" husbands: Men and masculinities in the Vietnamese transnational family. *Gender and Society, 25*(6), 717–739. https://doi.org/10.1177/0891243211430636

Hultman, M., & Pulé, P. M. (2018). *Ecological masculinities: Theoretical foundations and practical guidance*. London: Routledge.

Hussein, K., & Nelson, J. (1998). *Sustainable livelihoods and livelihood diversification*. IDS Working Papers No. 69. IDS. www.ids.ac.uk/publications/sustainable-livelihoods-and-livelihood-diversification/

Huynh, P. T. A., & Resurreccion, B. (2014). Women's differentiated vulnerability and adaptations to climate-related agricultural water scarcity in rural Central Vietnam. *Climate and Development*, 6(3), 226–237. https://doi.org/10.1080/17565529.2014.886989

IAITPTF. (2005). *Our knowledge for our survival: Regional case studies on traditional forest related knowledge and the implementation of related international commitments*. International Alliance of Indigenous and Tribal Peoples of Tropical Forests. www.international-alliance.org/publications.htm

Imai, K., Gaiha, R., & Kang, W. (2007). Poverty, inequality and ethnic minorities in Vietnam. *International Review of Applied Economics*, 25(3), 249–282. BWPI Working Paper 10. https://doi.org/10.1080/02692171.2010.483471

Indigenous and Tribal Peoples Convention. (1989). *76th ILC session, No. 169*. Geneva. www.ilo.org/dyn/normlex/en/f?p=NORMLEXPUB:12100:0::NO::P12100_ILO_CODE:C169

Institute of Development Studies (IDS). (2011). *Beyond women and girls' vulnerability: A debate on gender, climate change and disaster risk reduction*. https://portals.iucn.org/union/sites/union/files/doc/beyond_women_and_girls_vulnerability.pdf

Johnson, C. A. (2012). *Rules, norms and the pursuit of sustainable livelihoods*. IDS Working Papers No. 52. www.ids.ac.uk/publication/rules-norms-and-the-pursuit-of-sustainable-livelihoods

Jones, S., Saunders, J., & Smart, M. (2002). Repression of montagnards – conflicts over lands and religions in Vietnam's central highlands. *Human Rights Watch*. www.refworld.org/docid/45daf45c2.html

Kabeer, N. (2007). *Marriage, motherhood and masculinity in the global economy: Reconfigurations of personal and economic life*. IDS Working Paper No. 290. Institute of Development Studies. Brighton: University of Sussex. www.ids.ac.uk/files/dmfile/Wp290.pdf

Kahya, E., & Kalayci, S. (2003). Trend analysis of streamflow in Turkey. *Journal of Hydrology*, 289(1–4), 128–144. https://doi.org/10.1016/j.jhydrol.2003.11.006

Kryspin-Watson, J., Arkedis, J., & Zakout, W. (2006). *Mainstreaming hazard risk management in rural projects*. Disaster Risk Management Working Paper Series No. 13. World Bank. http://documents.worldbank.org/curated/en/602871468320732040/pdf/364620Mainstreaming0risk01PUBLIC1.pdf

Lambrou, Y., & Nelson, S. (2010). *Farmers in a changing climate: Does gender matter? Food security in Andhra Pradesh, India*. FAO. www.fao.org/3/i1721e/i1721e.pdf

Macchi, M., Oviedo, G., Gotheil, S., Cross, K., Boedhihartono, A., Wolfangel, C., & Howell, M. (2008). *Indigenous and traditional peoples and climate change: Vulnerability and adaptation*. Issues Paper IUCN. https://cmsdata.iucn.org/downloads/indigenous_peoples_climate_change.pdf

Makhabane, T. (2002). Promoting the role of women in sustainable energy development in Africa: Networking and capacity building. *Gender & Development*, 10(2), 84–91. https://doi.org/10.1080/13552070215909

Masika, R. (2002). *Gender, development, and climate change*. Oxford: Oxfam.

McElwee, P. (2010). *The social dimensions of adaptation of climate change in Vietnam*. Development and Climate Change Discussion Paper No. 17. World Bank. http://

documents.worldbank.org/curated/en/955101468326176513/The-social-dimensions-of-adaptation-of-climate-change-in-Vietnam

McLeod, E., Arora-Jonsson, S., Masuda, Y. J., Bruton-Adams, M., Emaurois, C. O., Gorong, B., Hudlow, C. J., James, R., Kuhlken, H., Masike-Liri, B., Musrasrik-Carl, E., Otzelberger, A., Relang, K., Reyuw, B., Sigrah, B., Stinnett, C., Tellei, J., & Whitford, L. (2018). Raising the voices of Pacific Island women to inform climate adaptation policies. *Marine Policy, 93,* 178–185. https://doi.org/10.1016/j.marpol.2018.03.011

MenEngage Alliance. (undated). *Men, masculinities and climate change: A discussion paper.* http://menengage.org/wp-content/uploads/2016/04/Men-Masculinities-and-Climate-Change-FINAL.pdf

MenEngage Alliance, & UN Women. (2014). *Men, masculinities, and changing power: A discussion paper on engaging men in gender equality from Beijing 1995 to 2015.* MenEngage Alliance. http://menengage.org/wp-content/uploads/2014/11/Beijing-20-Men-Masculinities-and-Changing-Power-MenEngage-2014.pdf

Minority Rights Group International. (2008). *Minority and indigenous groups.* Global Policy Forum (GPF). www.globalpolicy.org/component/content/article/171/29874.html

Momsen, J. (2010). *Gender and development.* London: Routledge.

Namey, E., Guest, G., Thairu, L., & Johnson, L. (2008). Data reduction techniques for large qualitative data sets. In *Handbook for team-based qualitative research.* Lanham: Rowman Altamira.

Nelson, V., Meadows, K., Cannon, T., Morton, J., & Martin, A. (2002). Uncertain predictions, invisible impacts, and the need to mainstream gender in climate change adaptations. *Gender and Development, 10*(2), 51–59. https://doi.org/10.1080/13552070215911

Oxfam. (2007). *Drought management considerations for climate change adaptation: Focus on the Mekong region.* An Interim Report. Vietnam. http://policy-practice.oxfam.org.uk/publications/drought-management-considerations-for-climate-change-adaptation-focus-on-the-me-112526

Oxfam. (2008). *Vietnam: Climate change, adaptation and poor people.* https://policy-practice.oxfam.org.uk/publications/vietnam-climate-change-adaptation-and-poor-people-112506

Oxfam, & UN-Vietnam. (2009). *Responding to climate change in Vietnam: Opportunities for improving gender equality.* A Policy Discussion Paper. UN-Vietnam. www.un.org.vn/index.php?option=com_docman&task=doc_details&gid=110&Itemid=266&lang=en

Paavola, J. (2008). Livelihoods, vulnerability and adaptation to climate change in Morogoro, Tanzania. *Environmental Science & Policy, 11*(7), 642–654. http://doi.org/10.1016/j.envsci.2008.06.002

Parreñas, R. S. (2005). *Children of global migration: Transnational families and gendered woes.* Stanford: Stanford University Press.

Parry, M. L., Canziani, O. F., Palutikof, J. P., van der Linden, P. J., & Hanson, C. E. (Eds.). (2007). *AR4 climate change 2007: Impacts, adaptation and vulnerability.* Contribution of Working Group II to the Fourth Assessment Report of the Intergovernmental Panel on Climate Change, IPCC. Cambridge: Cambridge University Press.

Pham, P., Doneys, P., & Doane, D. L. (2016). Changing livelihoods, gender roles and gender hierarchies: The impact of climate, regulatory and socio-economic changes on women and men in a Co Tu community in Vietnam. *Women's Studies International Forum (January–February 2016), 54,* 48–56. https://doi.org/10.1016/j.wsif.2015.10.001

Plant, R. (2002). *Indigenous peoples/ethnic minorities and poverty reduction: Regional report.* Asian Development Bank. https://think-asia.org/handle/11540/2968

Rashid, S. F., & Michaud, S. (2000). Female adolescents and their sexuality: Notion of honour, shame, purity and pollution during the floods. *Disasters, 24*(1), 54–70. https://doi.org/10.1111/1467-7717.00131

Salemink, O. (2003). *The ethnography of Vietnam's central highlanders: A historical contextualization, 1850–1990*. London: Routledge.

Sellers, S. (2018). *Climate change and gender in Canada: A review*. Women's Environment and Development Organization (WEDO). https://wedo.org/wp-content/uploads/2018/04/GGCA-CA-RP-07.pdf

Sujakhu, N. M., Ranjitkar, S., He, J., Schmidt-Vogt, D., Su, Y., & Xu, J. (2019). Assessing the livelihood vulnerability of rural indigenous households to climate changes in Central Nepal, Himalaya. *Sustainability 2019, 11*(10), 2977. https://doi.org/10.3390/su11102977

Swinkels, R., & Turk, C. (2006, September 28). *Explaining ethnic minority poverty in Vietnam: A summary of recent trends and current challenges*. Draft Background Paper for CEM/MPI Meeting on Ethnic Minority Poverty. Washington, DC: World Bank.

Tirado, M. C., Clarke, R., Jaykus, L. A., McQuatters-Gollop, A., & Franke, J. M. (2010). Climate change and food safety: A review. *Food Research International, 43*(7), 1745–1765. https://doi.org/10.1016/j.foodres.2010.07.003

Toulmin, C. (1992). *Cattle, women and wells: Managing household survival in the Sahel*. Oxford: Clarendon Press.

Tran, M. H. (2009, February 10–12). *Food security and sustainable agriculture in Vietnam*. Country Report. Vietnam. Submitted to the Fourth Session of the Technical Committee of APCAEM, Chiang Rai, Thailand. www.un-csam.org/Activities%20Files/A0902/vn-p.pdf

Tran, T. M. D., Hoang, X. D., & Do, H. (2006). *Định kiến và Phân biệt đối xử theo Giới [Prejudice and gender discrimination]*. Hànội: Nhà xuất bản đại học quốc gia Hànội.

Tschakert, P. (2014, August 18–22). Petra Tschakert on gender and climate change. *Hypatia*. Online Forum Interview. www.youtube.com/watch?v=ObCcQZBi6TU

Turner, B. L., Kasperson, R. E., Matson, P. A., McCarthy, J. J., Corell, R. W., Christensen, L., Eckley, N., Kasperson, J. X., Luers, A., Martello, M. L., Polsky, C., Pulsipher, A., & Schiller, A. (2003). A framework for vulnerability analysis in sustainability science. *Proceedings of the National Academy of Sciences, 100*(14), 8074–8079. https://doi.org/10.1073/pnas.1231335100

United Nations Development Program (UNDP). (2004, January 19–21). *The concept of indigenous people*. Background Paper Prepared by the Secretariat of the Permanent Forum on Indigenous People. New York: UNDP.

United Nations Environment Program (UNEP). (2016). *Global gender and environmental outlook*. Nairobi, Kenya: UN Environment. http://web.unep.org/ggeo

UN-Vietnam. (2009). *Vietnam and climate change: A discussion paper on policies for sustainable human development*. United Nations Vietnam. www.un.org.vn/en/feature-articles-press-centre-submenu-252/1020-viet-nam-and-climate-change-a-discussion-paper-on-policies-for-sustainable-human-%20development-viet-nam-and-climate-change-a-discussion-paper-on-policies-for-sustainable-%20human-development.html

UN Women. (2017). *Figures on ethnic minority women and men in Vietnam 2015: Based on the results of the survey on socio-economic situation of minority groups in Viet Nam 2015*. UN Women. https://www2.unwomen.org/-/media/field%20office%20eseasia/docs/publications/2018/01/figures-on-ethnic-minority-women-and-men-in-vietnam-ens.pdf?la=en&vs=1244

UN Women Watch. (2009). *Fact sheet: Women, gender, equality and climate change*. www.wocan.org/resources/women-gender-equality-and-climate-change-un-women-watch-fact-sheet#

Van Aelst, V. K., & Holvoet, N. (2016). Intersections of gender and marital status in accessing climate change adaptation: Evidence from rural Tanzania. *World Development, 79*, 40–50. https://doi.org/10.1016/j.worlddev.2015.11.003

Van Huynh, C., van Scheltinga, C. T., Pham, T. H., Duong, N. Q., Tran, P. T., Nguyen, L. H. K., Pham, T. G., Nguyen, N. B., & Timmerman, J. (2019). Drought and conflicts at the local level: Establishing a water sharing mechanism for the Summer-Autumn rice production in Central Vietnam. *International Soil and Water Conservation Research, 7*(4), 362–375. https://doi.org/10.1016/j.iswcr.2019.07.001

Van Leeuwen, L. (1998). *Approaches of successful merging of indigenous forest-related knowledge with formal forest management: How can modern science and traditions join hands for sustainable forest management?* Wageningen: National Reference Centre for Nature Management, Ministry of Agriculture, Nature Management and Fisheries.

Vinyeta, K., Powys White, K., & Lynn, K. (2015). *Climate change through an intersectional lens: Gendered vulnerability and resilience in indigenous communities in the United States.* PNW-GTR-923. U.S. Department of Agriculture, Forest Service, & Pacific Research Station. www.fs.fed.us/pnw/pubs/pnw_gtr923.pdf

Vo, T. H. (2012). *Assessment of long-term change of land-use in relation to climate and development conditions: A case study for the coastal district of Quang Nam Province, Vietnam* (Master's thesis), Asian Institute of Technology, Bangkok.

Vuong, X. T. (2008). Forest land allocation in mountainous areas of Vietnam: An anthropological view. In S. Robertson & T. H. Nghi (Eds.), *Proceedings of the forest land allocation forum on 28 May 2008* (pp. 45–55). Hanoi: Tropenbos International Vietnam in Cooperation with the Forest Protection Department with Support of MARD.

Walby, S. (1990). *Theorising patriarchy.* Oxford: Basil Blackwell. https://doi.org/10.1177/0038038589023002004

WEDO. (2007). *Changing the climate: Why women's perspectives matter.* https://wedo.org/changing-the-climate-why-womens-perspectives-matter/

Wetherell, M., & Edley, N. (1995). A discursive psychological framework for analysing men and masculinities. *Psychology of Men & Masculinities, 15*(4), 355–364. https://doi.org/10.1037/a0037148

World Commission on Environment and Development (WCED). (1987). *Our common future: Report of the world commission on environment and development.* Brundtland Report. WCED. https://sustainabledevelopment.un.org/content/documents/5987our-common-future.pdf

Writenet. (2002, January 1). *Vietnam: Indigenous minority groups in the Central Highlands.* Writenet Paper No. 5/2001. www.refworld.org/docid/3c6a48474.html

2 Impact of a changing climate on an ethnic minority community in a remote mountainous region of Viet Nam

The Co Tu ethnic minority of Central Viet Nam

The ethnic composition of the population of Ca Dy Commune, Nam Giang District, Quang Nam Province in recent years has consisted primarily of two main ethnic groups: the Co Tu and the Kinh. The Co Tu ethnic minority is the most prevalent group, accounting for approximately 86% of the commune's population. The Kinh (ethnic majority group) make up more than 11% of the commune's population, and the remaining part of the population comprises other ethnic minority groups (Ca Dy Commune People's Committee, 2013).

The Co Tu ethnic group is one of 54 officially recognized ethnic groups in Viet Nam and is one of the biggest ethnic minority groups in Central Viet Nam with 70,872 people, according to the General Statistics Office (GSO, 2015). The Co Tu population speaks Katuic, which belongs to the Mon-Khmer sub-group in the Austro-Asiatic language family. Their main locations are the mountains and natural forests of Quang Nam and Thua Thien-Hue Provinces in the Hien, Nam Giang, Nam Dong, and A Luoi Districts (they are also located in Xekong Province in Lao PDR, including along the border with Viet Nam's Quang Nam and Thua Thien-Hue Provinces). It is believed that they are one of the oldest ethnic groups in Viet Nam and that they were related at some point to the Cham and Kinh ethnic groups (T. V. Tran, 2009).

The Co Tu population is often described as small and extremely hardy (strong and resilient). This might be explained by their restricted diet and arduous lifestyle. During the French and American wars, they were regarded as keen allies of the North Vietnamese forces. They were universally respected for their sheer tenacity in battle and their knowledge of mountain survival techniques. They were also well known for being masters in hunting, and for that reason have been called "ghosts of the forests."

An interesting account of Co Tu involvement in the French and American wars is given in Arhem (2009, pp. 12–19).[1] According to Co Tu elders in Quang Nam Province interviewed for that study, the local Co Tu people found that the French, US and South Vietnamese forces during the prewar and war years treated them as inferiors ("savages"), whereas the North Vietnamese forces treated them as equals and provided them with "three-moon rice" and other benefits for their support and partnership in the war effort. In some Co Tu areas, including Ca Dy Commune, it is said that they suffered from exposure to Agent Orange (and therefore the highly toxic chemical contaminant dioxin) because the Ho Chi Minh Trail passed through their territories, and some of the Co Tu elders interviewed for that study (as part of the Katuic Ethnography Project) stated that after the US forces set up a "free fire zone" in their areas, they were given little or no time to evacuate before bombings destroyed their homes, fields and villages. For this and other reasons, they said that the Co Tu were loyal to the forces from the North and were able to maintain their own culture throughout this period, as well as appreciate their Vietnamese nationality; in fact, to this day, Ho Chi Minh is said to be venerated for his attitudes toward ethnic minorities, and many Co Tu villagers still have Ho Chi Minh's image on their home ritual altars. Based on interviews carried out for the present study, it appears that local villagers, particularly members of the older generation, continue to hold these perspectives.

Regarding gender relations, one of the most prominent characteristics of the Co Tu is their deeply patriarchal social system. Men are traditionally designated as the head of the household and hold political power in making decisions with respect to both the household and the community. Women, on the other hand, are responsible for looking after the family and for doing so-called light work (or "small work"), which will be discussed in detail in this and the following chapters. Culturally and socially, Co Tu women have always been considered subordinate to men, especially with regard to decision-making. Nonetheless, they play a key role in productive activities and are fully in charge of reproductive tasks.

In-depth interviews (IDIs), key informant interviews (KIIs) and secondary data regarding Co Tu cultural norms have revealed that both of the two main reasons given for women's lower status originated in the cultural belief that a daughter in the family is an embodiment of a debt between households. Regarding the first reason, interviewees noted the traditional system whereby a poor household could borrow money from a better-off household with a promise that, in the future, their daughter would marry the better-off household's son as a payment for the debt (and in most cases, they would keep their promise). Daughters were thus considered to be one of the family's main assets that could be "traded" to pay off debts. This custom is no longer as popular as it was in the past, but it persists as one of the two main explanations for the low status of women in the community that continues to the present time.

The second reason has to do with a debt incurred at the time of marriage. The idea of women as a family "asset" is reflected in the fact that, according to one female respondent, poor households preferred to have daughters because they could help their mothers with housework, and, more importantly, a mother would not have to worry about having enough resources to get her sons married. This is because weddings are very costly for Co Tu families, especially for the groom's family members, who often go into debt in order to fund the event. Before the wedding, the bride's parents can ask for a bride price from the groom's family.[2] This payment can consist of anything that they might be lacking at that moment – in most cases, the bride's family will ask for new clothes, rattan bags for carrying goods, a pig or buffalo for the wedding party, alcohol, rice, food and other items. One of the respondents stated that her parents asked for a bed as a payment for her wedding. Apart from the goods given to the bride's family, the groom's family must hold a large celebration and invite everyone in the community to drink and eat for two to three days. A wedding costs the groom's family a sizable amount and often adds to the burden of debt the groom's family faces. Once married, the young woman's work is compounded by her new position in the household of her husband's extended family and her responsibilities as a wife and mother, responsibilities that often include working to pay off her husband's family's debts.

In addition to the explanations for women's lower status cited earlier, another reason was simply that, as a forest-dwelling community in an isolated mountainous region, the Co Tu people had to deal regularly with attacks from dangerous animals, natural disasters, and other physical threats to their welfare. These threats required a response backed by physical strength, and men's traditionally designated roles as protectors of the community appear to have contributed – along with other specific cultural and historical factors – to raising their public status relative to women in the community.

In terms of religion and festivals, the Co Tu worship Giang (sometimes translated as "genie"), and it is said that every house has a special altar dedicated to Giang where they place horns and the heads of animals from earlier hunts. The Co Tu community has three big festivals: the buffalo stabbing festival (Đâm Trâu), the crop festival, and the ghost festival. During the buffalo stabbing festival, community members are said to worship the God of the Forest and their ancestors; they pray for good crops, health and other needs. The festival happens each year on September 2, and traditionally the village patriarch is the first one to kill the buffalo. The crop festival is held once the crops are mature enough to be harvested, and during this festival people pray to the God of the Rice Fields for better crops in the coming years. Villagers organize a large meal in which everyone is invited to join, and it is said that in the "old days" the large animals taken in the hunt were shared with all the villagers and that food was given particularly to the patriarchs and elderly; however, in current times, such sharing practices are no longer very common, as will be discussed in this and the following chapters. Finally, the ghost festival happens every two to three years, wherein people worship (honor) the ghosts of the forest and ghosts under their beds, who they traditionally believe are their ancestors (Luu, 2007).

According to T. V. Tran (2009), the Co Tu have strong animistic beliefs in the "spiritual essence of all things." They have deeply ingrained knowledge and a strong cultural and religious appreciation of the forest and its offerings. Forests have shaped the Co Tu villagers' way of life, and many customary laws and indigenous knowledge of the Co Tu people originated in and were developed through their practices of forest management (Arhem, 2009). Not only do their livelihoods depend on forests, but the forest also plays a major role in their spiritual, social, and cultural lives.

Impacts of sedentarization programs and other regulatory changes on the Co Tu community

Co Tu villagers traditionally practiced swidden (shifting, or slash and burn) cultivation based on a variety of highland crop cycles, centering on upland rice. Besides upland (also known as hill or dry) rice, they also plant corn, beans, cassava, sweet potatoes, bananas, and other fruits and vegetables. Hunting and the gathering of non-timber forest products (NTFPs) have also been crucial livelihood activities for the Co Tu. Apart from the main agricultural activities, in recent years nearly 95% of local households have been involved in home gardening for their own use, cultivating some vegetables and fruits; this form of livelihood, however, does not contribute much to a household's income and food needs (from KIIs with government officials and the household survey carried out in 2012 for this study). In sum, most of the Co Tu villagers have depended for generations on available natural resources in nearby forests and fields for subsistence purposes (Epprecht et al., 2011; Imai et al., 2007; T. V. Tran, 2009).

The Co Tu traditionally divide land and forest into two regimes: common and private property. Common property is for the community, clan or clan branch. Private property includes land for swidden cultivation, gardens and residences. Forests that can be used for exploitation are distributed by the villages, specifically by the village patriarchs. The distribution is also based on an agreement among the clans within a village, among clan branches (or lineages) within a clan, and among households within a lineage. The exchange, inheritance and transfer of land and forests usually take place within a clan (T. V. Tran, 2009). In the past, sites for the gathering of NTFPs could also be claimed by those who "found" them; others could not make use of the site once it was marked by the designated finders of the NTFPs in that specific location (Salemink, 2003; T. V. Tran, 2009).

For decades, the Government of Viet Nam has been trying to transform the lives of its ethnic minorities. It introduced sedentarization programs, including the Fixed Cultivation and Settlement program, which was implemented to give ethnic minorities a more "stable" lifestyle and access to better education, health care, and living conditions. More importantly, replacing traditional (often considered "backward") agricultural practices with more modern ones to avoid deforestation has been considered an essential part of the program (Plant, 2002). However, contrary to common assumptions, many ethnic minority groups do have a bounded territory and elaborate systems for delineating clan or community land (Salemink, 2003), and many observers have questioned whether traditional practices contributed at all to deforestation. What is clear is that as a consequence of the changing farming and management practices, the Co Tu are losing many of their customary laws and cultural practices. The Forest Land Allocation (FLA) program (discussed in more detail in Chapter 3), which distributed forestland to individual households, should therefore be seen within this context of sedentarization policies and programs (Bayrak et al., 2013).

Before the FLA program, forests in Viet Nam were, in principle, owned and managed by state-owned forest enterprises. However, in practice, there was an "open access" arrangement regarding the forests, which meant that many ethnic minorities were able to practice swidden agriculture, hunt, trap, and collect NTFPs freely in the natural forests. Their cultivated fields and forests were thus intimately connected and not clearly differentiated from one another.

The lives of those dependent on the forests have now been very significantly affected by the implementation of the FLA program. Not only did their livelihoods change, but their social, cultural and spiritual lives have also been affected. The Co Tu of Viet Nam have lost many traditional activities and customs due to the implementation of both the Law on Forest Protection and the FLA program. This has also weakened the role of the village patriarch: Villagers started losing the habit of consulting the village patriarch regarding production techniques and experiences, problems and conflicts that arise in the local area, whereas in the past, conflicts within villages or between villages were resolved by village patriarchs through negotiations and punishments. In addition to village patriarchs, the village elderly and unofficial diplomats were also traditionally involved in resolving conflicts (T. V. Tran, 2009). Every villager was expected to abide by the rules and obligations of traditional practices regarding natural-resource utilization and management, with rules being enforced both within and between villages (M. H. Tran, 2009). However, most of the conflicts and problems in the villages are no longer resolved by the village patriarch but rather are decided by local government officials, who often make top-down decisions (Bayrak et al., 2013; Plant, 2002).

In fact, criticisms of the banning of swidden cultivation and hunting activities are longstanding, as are criticisms of centralized policies such as "one size fits all" sedentarization and related policies that do not take local values, social conditions (including low population density) and specific features of the local natural environment into account.[3] Regarding the Co Tu community, Arhem and Binh (2006) state that the sedentarization program is "arguably the main factor responsible for the current predicament of the indigenous minority population in the region – shortage of swidden land, decline of shifting cultivation and depletion of game and fish in the vicinity of the relocated, permanent villages" (p. 42). They argue that the program reflects a misunderstanding about swidden agriculture and hunting and suggest that it is better to sustain them both in more traditional forms for environmental and sociocultural reasons.[4] They find that the program's restrictions on swidden cultivation and traditional hunting practices have actually led to worsening conditions for local communities rather than poverty alleviation in a meaningful sense; specifically, they cite a sharp increase in illegal logging and hunting on the part of individuals interested in personal gain rather than as an expression of group solidarity, which they say has resulted in greater depletion of wildlife and forests, particularly when outsiders join in to exploit local resources to sell in more distant markets.

Moreover, they note, "Elders and commune officials in these districts commented that this money [earned from hunting] was often spent on alcohol and entertainment in the district town, rather than being used for family or household consumption" (p. 66) – in other words, they indicate that it was often spent on men's personal consumption rather than for sustaining the household or community. As a consequence, they find that these policies have resulted in increased inequalities along with worsening diets overall (less game and fish than in the past) – particularly, but not only, in the case of those unable to work, who depended on sources of protein that had been shared with community members in the past but are now instead sold in the market or kept for the consumption of individual families. They further argue that the community's loss of self-sufficient production and the increased commercialization of agriculture has led to land disputes and over-exploitation of the land for commercial crops, and that this has been accompanied by the arrival of a larger Kinh population and the Kinh educational system, with the resulting loss of local community solidarity, cultural integrity and self-esteem. With this in mind, Arhem and Binh proposed recommendations that this study will come back to in Chapter 5. Unlike the present study, Arhem and Binh did not discuss climate change or the combination of forces that have caused sudden changes in gender roles, relations and hierarchies, but their observations from fieldwork carried out in the mid-2000s provide valuable perspectives on debates regarding policies that affect ethnic minority populations in the region.

We will now turn to a consideration of the impact of climate change on the Co Tu community. This will lay the groundwork for our discussion of serious climate threats and their implications for major changes in gender roles, relations and hierarchies in this and the chapters that follow.

Problems and vulnerabilities in the Co Tu community associated with climate change

For the Co Tu of Ca Dy Commune, the transitions discussed previously have been difficult, as the community continues to move from a relatively stable position of self-sufficiency to one of dependence on a very new physical, economic and social world that surrounds and defines them. This world is shaped not only by the policy changes discussed earlier, but in addition – and very significantly – by unpredictable climate conditions, along with unstable markets and rapid and destabilizing social and regulatory change.

In the case of this rural, poor and remote commune, the first and foremost concerns regarding the impacts of climate change relate to both food and livelihood security, as access to adequate food and sources of livelihood have come under threat due to serious climate trends over time. In the following chapters, we will see that due to gender-differentiated responsibilities regarding food-related and livelihood activities, Co Tu women and men have been affected in very different ways by the sudden and wide-ranging changes in climate and society.

Climate change in Ca Dy Commune, Quang Nam Province

The selected area of the present study is one of eight ecological regions of Viet Nam stretching from the coastal area in the east to the mountainous area in the west. According to Oxfam (2008), this region is prone to natural disasters, including floods, storms and drought, on an annual basis. The region is said to suffer from more climate extremes than are found in other parts of the country.

Regarding Quang Nam Province: With geographic coordinates of 14°57'10" to 16° 03'50" north latitude and 107°12'40" to 108°44'20" east longitude, Quang Nam is bordered by Thua Thien-Hue Province and Da Nang City in the north, Quang Ngai Province in the south, Kon Tum Province and Lao PDR in the west, and the East Sea in the east. With 1,057,474 ha (hectares), the province has two cities (Tam Ky and Hoi An), 16 districts and 241 communal official units. The population was reported as 1,494,000 in 2017 (Quang Nam Statistical Yearbook, 2018). The province is classified as one of the eight South Central Coast provinces, but it shares many of its social and environmental characteristics with the Central Highlands region just below its southern border.

Regarding the specific commune selected for study within Quang Nam Province, Ca Dy Commune is located in Nam Giang District (15°42'N, 107°47'E, elevation 173 ft.).[5] As a forested mountainous commune, Ca Dy Commune is characterized by many high mountains, valleys, and streams. The average altitude of the commune is about 300–500 m, with the highest points ranging from 800 to 1,000 m. The commune's land is divided into five slope classes, ranging from class I (0–2.5° slope), where the villages are situated, through class II (2.6–5° slope), where the hill farms are located. The topography continues up to class V (> 20.1°), which is mainly forest. The field areas of villagers are mostly located on class II hillside slopes (ADB, 2010).

Apart from the fact that it is threatened by climate risks, Ca Dy Commune was selected for the study because it has experienced socioeconomic changes brought about by the construction of a new national road – National Route 14B – that was begun in 2004 and completed in 2005. Ca Dy is located on the main route that extends from the north to the south of Viet Nam. The opening of the road has brought along opportunities for livelihood improvement to the community but at the same time has helped create serious challenges to traditional livelihoods, practices and beliefs.

Local climate conditions

The climate of Ca Dy Commune is characterized by tropical monsoons in valleys with two seasons. The rainy season, from August to December, has an average rainfall of 2,500 mm and an average of 85% humidity. The dry season is from January to July, with an average rainfall of 50–70 mm. Rainfall is unevenly distributed geographically, with approximately 2,000 mm in the coastal area and 3,000–4,000 mm in the upland part of the province, and is more uneven in terms of the timing of rainfall, with very heavy rains in some months and very little in others (Ho & Umitsu, 2011; Quang Nam Statistical Yearbooks, 2014, 2017).

There are two main types of wind: The first is the southwest wind in June and July, lasting for a long duration and accompanied by high temperatures, which can have significant impacts on agricultural productivity, especially on rice crops (this wind is locally called a "Lao wind," and is also known as a Foehn wind; it is a type of dry, warm, downslope wind that occurs on the leeward side of a mountain range). The second is the northeastern wind that extends from October to December; it brings heavy rainfall and low temperatures, thus causing negative impacts on agricultural production and livestock raising, given that many animals cannot survive for long in these unusually cold conditions (ADB, 2010).

The agricultural sector in Quang Nam Province is very important because it meets nearly 100% of provincial food needs (Quang Nam Statistics Office, 2014). The most recent land survey of Ca Dy Commune is summarized in Table 2.1.

As noted, the total forested area is 15,053 ha, of which 9,513 ha is natural forest and 5,540 ha is planted forest. Due to its ecological characteristics, the natural forest occupies the largest part of the commune's total natural land area (47.56%), whereas cultivated land covers a much smaller part (2.46%) of the commune's natural landholdings. In recent years, of the total forested area 100 ha has been assigned to households, 1,183 ha to the Forest Management Board, and the remainder to the Commune People's Committee as a result of the regulatory changes discussed in more detail in Chapter 3 (ADB, 2010).

NTFPs found in this commune are mainly rattan, other vegetables and fruits, nuts, tubers and honey. Once the Co Tu women started learning how and where to sell their surplus agricultural products and NTFPs, two major sources of cash income became malva nut and rattan, which were collected from the forests by both men and women. Rattan was a valuable forest product for the local community: First, it was used to make baskets, which were considered one of the most important items owned by Co Tu families; it is also a Co Tu symbol that is associated with Co Tu women as they used them to carry food, firewood, harvested rice and whatever else was needed. Rattan has thus been a main NTFP for local villagers, but in recent years its availability has declined so much that villagers estimated during focus group discussions (FGDs) that it is now is only about 30% of what it was 10–20 years ago. During FGDs, commune members agreed that rattan had not only become scarce, but its quality had also decreased because of over-exploitation, rather than from adverse weather conditions alone. It is said to be difficult now to find rattan the same length as it was in the past (this is important because rattan's length is what determines its price).

Malva nut, or the so-called Uoi fruit, is a species of tree native to mainland Southeast Asia. Its seeds are used in traditional Chinese medicine either as a

Table 2.1 Land use status in Ca Dy Commune

Land use type	Coverage area (ha)	Percentage (%)
Total natural area	20,000	100
Agricultural land	492	2.46
Forested land	15,053	75.27
Natural forest	9,513	47.56
Planted forest	5,540	27.7
Other land	4,455	22.28

Source: Ca Dy Commune People's Committee, 2013

coolant to soothe the throat or as a cure for gastrointestinal disorders; it is also used as a beverage. Malva nut can be harvested every three to four years. In recent years, increasing demand from Chinese buyers has made its price rise rapidly, which has led to destructive harvesting techniques. Twenty years ago, NTFP collectors waited for the nuts to ripen and collected them after they had fallen and scattered away from the tree; however, over the last few years, because of its high profitability, collectors (including a large number of non–Co Tu collectors) chop down the tree to reach the nuts while they are still on the tree to save time and maximize the volume of nuts obtained. As a result, the quantity of the local malva nut available for use has declined significantly.

Other NTFPs include langsat fruit (Bon Bon), bamboo, and grass for thatching, and these were often gathered together with firewood and timber. However, these too have either suffered due to reduced demand and low prices (langsat fruit and thatching grass) or because access has been greatly restricted, rather than climate change being a clear causal factor. It is important to note that the impact of the changing climate on these specific forest-based resources is not yet well understood, but a combination of factors undoubtedly has led to a decline in the harvesting of NTFPs in this region.

The changing climate and people's perceptions

We will now turn to both local perceptions and statistical meteorological evidence regarding the impact of climate change on this research site. Comparisons between these two sources of data will serve the purpose of assessing changes in climate more reliably.

First, local perceptions of climate change will be presented by discussing key points drawn from IDIs, KIIs and FGDs as well as from observations during fieldwork. The main focus of the research is about the impressions and perceptions held by local Co Tu women and men regarding changes that have happened in their lives, with somewhat differing responses according to gender, most likely due to their gender-differentiated roles and experiences in the fields and forests.

Following this, climate change will be discussed statistically by using the MK test, analyzing meteorological data collected in the Tra My and Tam Ky meteorological stations for a period of almost 30 years, beginning in 1980. Each finding will be used to compare, explain and verify or contrast with local perceptions of climate variations. This will be followed by more recent data and projections into the future from provincial and district-level analyses. We will then compare what we know from these different sources of information regarding climate-related changes that have taken place in the area, particularly over the past two to three decades.

Evidence from primary data about local perceptions of climate change

Primary data collected from research fieldwork both in 2011–2012 and 2019 confirmed that over the last two to three decades, the Co Tu community in Ca Dy

Commune has been experiencing climate change in the forms of (1) unpredictable variations; (2) extreme weather changes, with increasingly hot summers and
increasingly cold winters; and (3) an increasing frequency of weather extremes
caused by prolonged hot and dry weather that can lead to dry spells and prolonged
heavy rainfall that can lead to serious flash floods. The interviews (IDIs and
FGDs) cited next were conducted during the 2011–2012 and 2019 visits, whereas
household survey results are drawn from the 2012 quantitative findings carried
out for this study.

The change most often mentioned during IDIs and FGDs was that the weather
has become very erratic and extreme. Co Tu elders stated that they believe the
weather no longer follows previous patterns. It now fluctuates unpredictably
every year, and rain, floods and storms come earlier or later than expected. For
this reason, they argue that the Co Tu villagers can no longer predict weather patterns for the purposes of agricultural production. As one 69-year-old man noted
during an IDI in one of the early visits to the community:

> The weather now is so unpredictable. Before, we'd know exactly when the rainy
> season would come and when the dry season occurs. We could also predict when
> would be the best time for planting seeds. I don't know why the weather has
> become so erratic. There could be one or two months with no rain at all and then
> heavy rains come, leading to flash floods because of the huge rainfall within a
> too-short period of time. Consequently, rice crops may be entirely destroyed.

Other respondents interviewed at this time supported this view:

> Rain has become increasingly rare. In some years, it was so rare that rice
> seed could not survive after being sown. For example, in 2008, we did not
> have rain at all for nearly two months, in March and April. We had a very
> bad crop that year.
>
> (FGD, 47-year-old man)

> Weather has become unpredictable, even at the same time of the year. In
> 2008, we had a terrible crop loss because there was almost no rain at all in
> March and April – whereas during the year after that, in 2009, we also had
> a very bad crop because of flash floods since in that year the rain was much
> heavier than it usually is.
>
> (FGD, 69-year-old man)

They also stated that the increasing frequency of flash floods and heavy rains,
particularly in a context of steep slopes, has caused soil degradation and led to a
reduction in rice yields, as noted in the following quote:

> The seasons are also changing a lot. The rainy season can come earlier
> or later. It's the same with the dry season. This extreme and unpredictable

weather, along with over-cultivation leading to soil degradation, causes crop loss and reductions in rice yields.

(KII with the head of Nam Giang District's
Commune People's Committee)

Based on the results of this study's household survey, Figure 2.1 illustrates people's perceptions about climatic factors that have been changing over the last two to three decades. This figure shows that the three climate-related changes mentioned most by respondents were: (1) the weather pattern has become more irregular (selected by 87% of respondents); (2) summers have become hotter (86%); and (3) there are increasingly prolonged hot and dry periods in the annual weather patterns (77%). The other changes included the increasing number of prolonged days of heavy rain, frequency of dry spells, increasing frequency of flash floods, and colder winters. The increase in the frequency of weather extremes was mentioned in the survey by 67% of the respondents.

Regarding the greatest change associated with irregular weather patterns, elders in the community clearly recalled a different pattern from their childhood, as indicated in a quote from one of the village patriarchs:

In recent years, the weather has become more severe than ever. It can be tremendously hot. Hot and dry weather could last for over a month, even two months, without any rain. This rarely happened when we were young. Rice and other crops have all died because of the lack of rain. When the weather is too hot, it also prevents rice from growing.

(IDI with the Pa Pang Village patriarch)

This quote emphasizes the effect of the hot and dry weather on the area's upland rice crop production, which is the most important crop for the Co Tu community. Much like prolonged hot and dry weather, prolonged periods of heavy rain were also repeatedly mentioned by many respondents because these extreme weather events have strong negative impacts on agricultural production.

In 2019, seven years after these initial interviews and the household survey were conducted, the perceptions regarding a changing climate were confirmed by both elderly and middle-aged women respondents, with "increasingly hot and dry weather" as the most commonly noted change (FGDs and IDIs). As in earlier interviews, respondents frequently used the terms "unpredictable," "abnormal," "irregular change," "extreme," "unbearably hot" and "little rain" in these recent interviews to describe new weather patterns.

It should be noted that the term "climate change" rarely appeared during interviews (IDIs and FGDs) with local villagers because most of these respondents were not used to differentiating between the terms "climate" and "weather." All of the factors related to temperature and/or rainfall were simply regarded as "weather." When asked about changes in climate that have occurred over the past 20–30 years, respondents promptly stated that the weather has gotten hotter. After

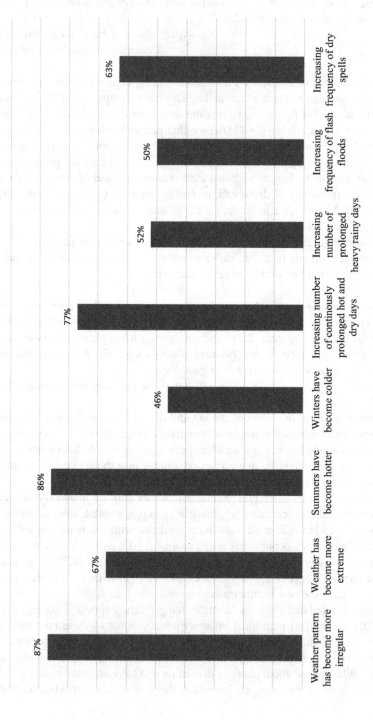

Figure 2.1 Changing climate-related factors based on people's perceptions

Source: Household survey, 2012, n=300

being asked to clarify whether "hotter" refers to weather in general or only to temperatures during the summer, they confirmed that summers have gotten hotter, but not the weather in general because in some years the winters were unbearably cold.

In contrast to local villagers, the terms "climate," "climate change" and "global warming" were mentioned a number of times by government officials, including officials who worked for the district's Department of Environment and Department of Agriculture, as well as the Commune People's Committee:

> *Yes, over the last twenty years the climate has changed. I think you must know that it's because of global warming. The temperature has increased. The weather has gotten hotter.*
>> *(KII with a government official working at the Department of Environment of Nam Giang District)*

> *Of course, climate change affects agricultural production. You know that the Co Tu cultivate upland farming. They leave their crops entirely to nature's care [without using other inputs, relying entirely on natural weather patterns].*
>> *(KII with a government official working at the Department of Agriculture of Nam Giang District)*

It is important to note that it is not only the characteristics and intensity of weather extremes that have negatively affected the local agricultural production systems. The *timing* of extreme weather is another important and more concerning factor that can have severe impacts on agriculture in general and the rice crop in particular, which is by far the most important crop for Co Tu families. Interviews indicated that the Co Tu villagers used to face similar weather events in the past. However, it is said that these events did not affect crop production as much as in recent decades; what has exacerbated the impact of extreme weather events, according to local perceptions, is the timing of those weather events. If flash floods came after rice harvesting, it would not have destroyed the crop totally or in part. This was also true of dry spells – respondents noted that, back in the 1980s, the hot and dry spells, if any, often came after April when the rice had grown, and before August, which is the flowering time for rice in this area (this comes a month or two before harvest time). However, according to Co Tu respondents, in recent years the climate events were unpredictable and "could come anytime and lead to bad crops," as described in the following quotes:

> *In 2000, we had a very bad crop because we did not have any rain during March. It was too hot and dry so the rice could not grow. Then in the following year, we had another crop loss, which was caused by flash floods that came during August or September. I don't remember the exact year, but I remember that I was not able to complete rice harvesting before the flood came, so it destroyed half of my crop.*
>> *(FGD, 47-year-old woman)*

> *Recently, farming has become so much more difficult than it used to be fif-teen or twenty years ago because of unfavorable weather. Hotter and drier weather during March and April has seriously affected rice growth. Flash floods that came earlier, when we had not yet finished the rice harvesting, caused crop loss for us. We faced food deficits [food shortages and hunger] because of changing weather patterns.*
>
> *(IDI with a 49-year-old woman)*

In addition, respondents said that during the last 20–30 years they had experi-enced abnormally low water levels several times. This was supported by an agri-cultural extension officer of the Department of Agriculture of Nam Giang District, who noted:

> *Climate change has caused irregular weather patterns. In some years, there was very little rain, and in some others, rains came in huge amounts.*

Apart from commenting on increasingly hot summers, almost half (46%) of the respondents surveyed in the first field visits noticed that winters in some years were colder than in the past. Although these winters were not perceived to be as damaging as increasingly hot summers, cold winters also worsened the poor liv-ing conditions in the community. A few respondents linked the cold weather with crop losses when it affected the growth of rice seeds:

> *The weather nowadays has become so much more unpredictable. Last year, and a few years before that, winters lasted for a very long time. The tempera-tures during winter were so low that rice seeds could not grow.*
>
> *(IDI with a 43-year-old man)*

The majority of respondents (94%) surveyed in the first visit agreed that over the previous two decades climate had changed, and more importantly these changes negatively affected their lives. Specifically, they found the changes to be very challenging and uncomfortable. Hot and dry weather not only made their liv-ing conditions harsher, it also made their farming activities more difficult. Long-lasting heavy rains and flash floods were not reported as causing serious damage to people's houses and assets, but they made harvesting difficult and sometimes impossible:

> *The weather now is very intolerable. In the summers, it could be so hot that we had to always use a fan [the weather was very hot and sticky]. It has been much hotter than it was in the past. In addition, the Lao wind in June and July made the summer heat become unbearable.*
>
> *(IDI with a 62-year-old woman)*

> *In comparison to the weather twenty years ago, it has become more uncom-fortable, either too hot or too cold. Rains last longer. So has the hot and dry*

weather. Prolonged heavy rains could also be dangerous because they create mountain landslides.[6]

(KII with the Head of Ro Village)

Gender and local perceptions

From a gender perspective, it was interesting to note that the Co Tu women and men had slightly different impressions about the changing climate. During the first period of data collection, out of 32 male IDI respondents, 31 stated that they believed hotter weather and irregular changes in weather patterns were the two most significant changes in climate, having the strongest impacts on their daily lives as well as on agricultural production. On the other hand, 30 out of 32 female respondents said during IDIs that they thought the irregular changes in weather patterns and increasing frequency of weather extremes would affect their agricultural production – their livelihood survival strategy – the most, with relatively little worry about the impact on their comfort and daily lives.

In this way, the Co Tu women and men agreed that climate has changed in irregular and unexpected ways; however, they had somewhat different opinions regarding the importance of hotter weather versus the increasing frequency of weather extremes. More male respondents (145 compared to 112 female respondents) noticed that the weather was getting hotter. Male respondents stated that the number of hot days without rain had risen significantly. As one male respondent who helps with some of the farm work argued, hot weather had made participating in farming activities more difficult for him:

> *I don't know why it has become so hot during summer in recent years. Sometimes I feel that working in the fields has become unbearable.*
>
> *(IDI with a 43-year-old man)*

Meanwhile, during IDIs, more women than men seemed to mention increasing extreme weather events rather than the discomfort of hot summer temperatures. This was not revealed through the household survey, but rather during interviews, when respondents discussed in detail their thinking about the changes in climate-related factors. Female respondents in both the 2011–2012 and 2019 interviews cited the increased frequency and intensity of hot and dry weather and the increasingly unpredictable weather extremes as the two most concerning problems and as having the greatest impacts on people's lives and livelihoods, particularly because of the effect on their agricultural production efforts and more specifically on the rice crop. The following is one such example, from an IDI with a 54-year-old woman:

> *In recent years, little rain with high temperatures during summers and long-lasting heavy rains and flash floods during the rainy season were the two weather extremes that caused us the most damage in terms of agricultural production, making our lives difficult because we had to deal with hunger.*

It is likely that the differences between Co Tu women's and men's concerns about changing climate factors could have originated from gender-differentiated livelihood tasks. As the primary agricultural laborers, the Co Tu women were more responsible for crop cultivation and would be concerned more about the most crop-threatening factors, such as unexpected flash floods and long-lasting heavy rains. They could easily lose a part or the entire crop because of flash floods and dry spells. More importantly, the Co Tu women have traditionally and socially been designated to be responsible for providing food for their families. They are the ones who worry the most when their children go hungry. That is likely to be why weather extremes and unexpected timings were mentioned more in the women's responses.

In contrast, Co Tu men were traditionally in charge of hunting, trapping and cutting big trees for a new crop, activities that were normally conducted in the shade rather than during the full heat of the day. Therefore, the Co Tu men who helped to varying extents with farm work strongly expressed their concern about the increasingly hot weather because they no longer worked in forests where they were surrounded by big trees; in addition, they were not as used to working in the unprotected fields under conditions of increasing temperatures, as was true for women working in the fields. (These changes in gender roles and responsibilities will be discussed in greater detail in the next chapter.)

In sum, according to local villagers, climate has changed over the past two decades with regard to three major factors. First, the weather pattern of rainy and dry seasons has become *unpredictable*, with strongly negative impacts on crop cultivation. Second, there has been a consistent increase in the frequency and intensity of *hot and dry weather*, which can last for several months without rain, greatly affecting people's lives and livelihoods. Third, the frequency of *weather extremes*, including prolonged heavy rains, dry spells and flash floods, has increased with unpredictable timing. The changing climate has not only affected people's daily lives, but, more importantly, it has led to a significant increase in *food insecurity*, which will be taken up in the context of women's and men's very different responses to these new challenges in Chapter 3.

The changing climate: scientific evidence based on meteorological data

The respondents' perceptions about the changing climate correspond with findings from studies about climate change in Viet Nam in general and in Quang Nam Province in particular. As noted in Chapter 1, the assessment report on climate change (ISPONRE, 2009) shows that temperatures in Viet Nam have increased by 0.05–0.20°C in recent years, and the sea level has increased by 2–4 cm per decade over the last 50 years. According to a study conducted in 2011, the average air temperature in Quang Nam Province increased about 0.06–0.1°C per decade from 1980 to 2010 (Vo, 2012). The monthly rainfall in the province also had an upward trend with a standard deviation of 500–700 mm in most areas from 1991 to 2010, and the frequency of floods and droughts has tended to increase in recent

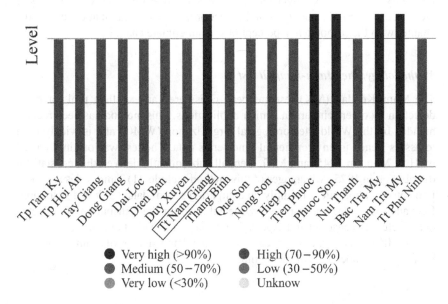

Very high (>90%) High (70–90%)
Medium (50–70%) Low (30–50%)
Very low (<30%) Unknow

Figure 2.2 Frequency of storms in the area under study (1960–2017)
Source: Viet Nam's Climate Index (UNDP, 2018)

years. Quang Nam is also prone to storms, as seen in Figure 2.2 in which Nam Giang District is classified as having one of the highest ("very high") frequencies of storms during the 1960–2017 period.

As a result, soil erosion has become more significant over the years. The quality of cultivated land has been affected, and this has caused a reduction in rice yields.

In the following section, secondary sources of hydrological data will be analyzed to evaluate the data and examine trends regarding climate change over the past 30 years. Quang Nam Province has eight meteorological stations in total; however, there is no meteorological station located at a commune level, thus the recorded meteorological data discussed in this section have been drawn from the district level. The district-level data were from the two closest meteorological stations that have available meteorological data records for almost 30 years: Tra My and Tam Ky stations (the Tam Ky meteorological station in particular is only 122 km from Ca Dy Commune, but both are relatively close to the study area).

Again, depending on the type of data (temperature, rainfall and others), it should be noted that the majority of data that are available to us from these two stations extend from 1980 to 2005 or 2008. We will first analyze these data and then use additional analyses of secondary sources of time series hydrological data, extending to 2017 and with projections to 2050, in order to examine trends regarding climate change over recent decades and beyond. Finally, the results of the meteorological data analysis in this section will be compared at the end of the chapter with the respondents' perceptions about climate change.

Methodology: the Mann-Kendall test

The Mann-Kendall (MK) test has been considered an excellent tool for trend detection by other scholars in similar applications. This method has been recommended by the World Meteorological Organization (WMO) and is widely used to assess trends in environmental time series data. The test was originally used by H.B. Mann in 1945, and M.G. Kendall subsequently derived the test statistic distribution in 1975. Advantages of this method include: (1) Time series are taken into consideration but not simply by comparing sets of data; (2) it is distribution-free; and (3) it does not assume any special form for the distribution of collected data, including censored and especially missing data (Yue et al., 2003).

The MK test is applicable in cases when the data values x_i of a time series can be assumed to obey the following model:

$$x_i = f(t_i) + \varepsilon_i,$$

where $f(t)$ is a continuous monotonic increasing or decreasing function of time, and the residuals e_i can be assumed to be from the same distribution with zero mean. It is therefore assumed that the variance of the distribution is constant in time.

The purpose is to test the null hypothesis of no trend, *H0* (i.e., the observations x_i are randomly ordered in time), against the alternative hypothesis, *H1*, where there is an increasing or decreasing monotonic trend. For time series with fewer than ten data points (n < 10), the S test is used, and for time series with ten or more data points (n > 10), the normal approximation is used.[7] The test, therefore, is completely suitable for use with the data collected from the two meteorological stations in Quang Nam Province with the application of normal approximation.

The statistically significant trend is tested by the Z statistic, calculated as follows:

$$Z = \begin{cases} \dfrac{s-1}{\sqrt{VAR(S)}} & \text{if } S > 0 \\[2mm] 0 & \text{if } S = 0 \\[2mm] \dfrac{s+1}{\sqrt{VAR(S)}} & \text{if } S < 0 \end{cases}$$

in which $VAR(S) = SD^2(S) = \dfrac{1}{18}\left[n(n-1)(2n+5) - \sum_{j=1}^{g} t_j (t_j - 1)(2t_j + 5) \right]$

where n is the sample size, g represents the number of groups of ties in the data set (if any) and t_j is the number of ties in the jth group of ties.

A positive (negative) value of Z indicates an upward (downward) trend. To examine whether the monotone trend is significant, the multitude of calculated Z statistic is compared against critical values of a standardized normal distribution corresponding with chosen significance levels of α, for example, α being 0.001, 0.01, and 0.05. If the Z statistic, at absolute value, is larger than the critical value, say 1.96 at $\alpha = 0.05$ of a two-tailed test, we reject the hypothesis *H0* (again, *H0* indicates "no trend"), meaning that there exists a significant trend in the observations. The significance level is displayed as:

*** if trend is significant at $\alpha = 0.001$
** if trend is significant at $\alpha = 0.01$
* if trend is significant at $\alpha = 0.05$

The level of significance $\alpha = 0.01$ means that there is a probability of 1% that we make a type I error in rejecting *H0* (that no trend exists). Thus, the significance level 0.01 means that the existence of a monotonic trend is very probable. The significance level 0.05 means that there is a 5% probability that we are mistaken when rejecting *H0*.

Temperatures: trends regarding annual mean temperatures

By employing the MK test to identify a possible trend in annual average (also called annual mean) temperatures (see Figure 2.3), we find that an increasing trend would be statistically significant at the 5% level of significance.

According to the results shown in Table 2.2, the calculated Z statistic is 1.83, and the probability of obtaining such a value is well below 5%, indicating that there is no clearly increasing trend at the 0.05 (5%) level of significance. From this test, it can be concluded that there is not enough evidence to confirm an increasing trend of annual mean temperatures in this area between 1980 and 2005 at a significance level of 5%.

For the data collected in Tam Ky station between 1980 and 2008, the results of the MK test are shown in Table 2.3. The calculated Z statistic is 1.22, indicating that, again, there is no clearly increasing trend of annual mean temperatures at the 0.05 (5%) level of significance, based on the data available over three decades in this province.

In both cases of data collected from different stations, the computed R^2 results using MS Excel are all small: 13.12% and 4.7%, respectively. Therefore, these results further support the conclusion that there is no significant trend regarding annual mean (average) temperatures in Quang Nam Province, whether increasing or decreasing, within these three decades. However, this conclusion again applies only to *annual* mean temperatures, not *monthly* mean temperature fluctuations, as will be discussed next.

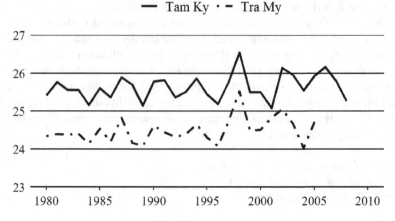

Figure 2.3 Annual average temperatures: Tam Ky and Tra My stations from 1980 to 2008
Source: Tam Ky and Tra My meteorological stations, Quang Nam Province

Table 2.2 Annual mean temperatures – Tra My station (MK test)

Time series	First year	Last Year	n	S statistic	Z statistic	p-value
Annual Mean Temperatures	1980	2005	26	84	1.83	0.067

Source: Tra My Meteorological Station, Quang Nam Province

Table 2.3 Annual mean temperatures – Tam Ky station (MK test)

Time series	First year	Last Year	n	Test S	Test Z	p-value
Annual Mean Temperatures	1980	2008	29	66	1.22	0.222

Source: Tam Ky meteorological station, Quang Nam Province

Monthly mean temperature fluctuations

Statistically, the hypothesis of possible trends in annual mean temperatures in Quang Nam Province has been proven wrong using data collected from both Tam Ky and Tra My stations. Nevertheless, it is necessary to look further by statistically comparing how the average temperatures of each month varied in recent decades. This would help us recognize possible negative impacts of climate change in this area. To conduct such an evaluation, a coefficient of variation (CV) for temperatures in each month was calculated from 1980 to 2008. The results are shown in Table 2.4. Calculation results showed that the CV values of annual mean temperature fluctuations are 1.37% (Tra My station) and 1.32% (Tam Ky station). This further supports the statement made earlier

Table 2.4 Coefficient of variation (CV) of monthly and yearly temperatures (CV%)

	Jan.	*Feb.*	*Mar.*	*Apr.*	*May*	*Jun.*	
Tra My	4.51	4.97	3.66	2.66	3.66	2.66	
Tam Ky	4.84	4.95	3.61	2.54	2.50	2.49	
	Jul.	*Aug.*	*Sep.*	*Oct.*	*Nov.*	*Dec.*	*Annual*
Tra My	2.14	2.86	1.19	2.63	3.77	4.95	1.37
Tam Ky	1.98	2.57	1.52	1.96	3.73	4.72	1.32

Source: Tra My and Tam Ky meteorological stations, Quang Nam Province

that claimed no statistically significant trend of annual air temperatures exists throughout the three recent decades.

Looking at CV values from different months, however, it is clear that the *annual* mean temperatures did not change much, but the *monthly* mean temperatures fluctuated more significantly. Also, the CV values of the winter months (December, January, February) are higher than those from the summer months (June, July, August), indicating that temperatures during the winter months fluctuated and varied widely.

Thus, while the annual mean temperatures have not changed, temperature fluctuations have varied widely from month to month. As a result, there were months during the summer when the monthly temperatures increased year after year (same month, different years), and there were also months (December, January and February) when the monthly temperatures decreased year after year (same month, different years). These findings support the statements of respondents that the summer season has been increasingly hotter, and winters have been increasingly colder. Harsher weather has not only made people's daily lives more uncomfortable and difficult, but more importantly, it has made agricultural production – their livelihood survival strategy – unpredictable and difficult to carry out.

Rainfall

According to data recorded in the Tam Ky station, annual rainfall in the research area varied significantly, from 1,500 mm/year to approximately 4,500 mm/year throughout recent decades, as shown in Figure 2.4. Meanwhile, as recorded in the Tra My station, annual rainfall amounts varied even more drastically, from about 2,500 mm/year (in 1982) to more than 7,000 mm/year (in 1996), as seen in Figure 2.4, and the MK test results for annual rainfall, rainfall in the rainy season and rainfall in the dry season for the two meteorological stations for the same period (1980–2008) are presented in Tables 2.5, 2.6 and 2.7, respectively.

Based on the data spanning almost three decades, the Z statistics of both Tam Ky and Tra My stations derived from annual rainfall data are all insignificant at the 5% level of significance. That is, the MK test results suggest there is no significantly increasing trend in annual rainfall, as recorded in both stations.

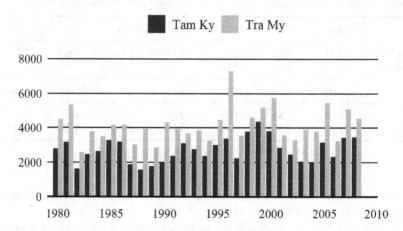

Figure 2.4 Annual rainfall (mm): Tam Ky and Tra My stations from 1980 to 2008

Source: Tam Ky and Tra My meteorological stations, Quang Nam Province

Table 2.5 Annual rainfall (MK test)

Station	First year	Last year	N	S statistic	Z statistic	p-value
Tam Ky	1980	2008	29	78	1.44	0.149
Tra My	1980	2008	29	48	0.88	0.378

Source: Tra My and Tam Ky meteorological stations, Quang Nam Province

Table 2.6 Rainfall in the rainy season (MK test)

Station	First year	Last year	N	S statistic	Z statistic	p-value
Tam Ky	1980	2008	28	34	0.65	0.514
Tra My	1980	2008	28	24	0.45	0.650

Source: Tra My and Tam Ky meteorological stations, Quang Nam Province

Table 2.7 Rainfall in the dry season (MK test)

Station	First year	Last year	N	S statistic	Z statistic	p-value
Tam Ky	1980	2008	29	82	1.52	0.129
Tra My	1980	2008	29	49	0.90	0.368

Source: Tra My and Tam Ky meteorological stations, Quang Nam Province

By employing the MK tests for seasonal rainfall data for the dry (March to September) and rainy (October to February) seasons in the research area, further analysis can be made regarding the possibilities of increasing/decreasing trends in rainfall as recorded by both stations. The results derived from these tables suggest there is not enough evidence to accept the hypothesis of a possible increasing/ decreasing trend in rainfall, not only on an annual scale but also on a seasonal scale (i.e., for rainy and dry seasons). Nevertheless, apart from annual or seasonal trends regarding rainfall, it is also interesting and necessary to look at how *monthly* data varied throughout recent decades. Obviously, patterns of rainfall are not only reflected by annual amounts but also by how rainfall was distributed between months. To understand these patterns and evaluate rainfall fluctuations, the CV values of months in each season, dry or rainy, are taken into account.

Rainy season rainfall trends

First, the fluctuations can be roughly assessed based on data on monthly rainfall in the two stations, shown in Figures 2.5 and 2.6. These charts suggest that the rainy season was actually concentrated mostly in the last three months of the year, namely October, November and December. For example, both Tra My and Tam Ky stations experienced low rainfall in November 1986, at about 300 mm, but in November 1996, Tam Ky station recorded 1,500 mm and Tra My station recorded 2,500 mm (i.e., five times and eight times higher than ten years earlier).

Table 2.8 shows a greater amount of rainfall within these three months, and the fluctuation of rainfall throughout these three decades was also much more noticeable and considerable, varying from 50% to 80%. To further clarify this fluctuation, the CV values of rainy months from both stations have been calculated and the results have been presented in this table.

The combination of both a high level of fluctuation and a greater amount of rainfall within these three months suggests that the rainfall factor in the research area within the rainy season has been very unstable year by year, making it difficult to forecast. This also explains why eight floods have been recorded over the

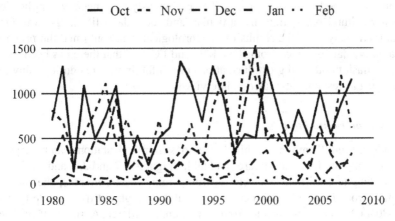

Figure 2.5 Monthly rainfall (mm) in the rainy season: Tam Ky station from 1980 to 2008

Source: Tam Ky meteorological station, Quang Nam Province

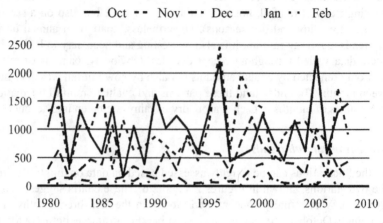

Figure 2.6 Monthly rainfall (mm) in the rainy season: Tra My station from 1980 to 2008
Source: Tra My meteorological station, Quang Nam Province

Table 2.8 CV values of months in the rainy season (CV%)

Station	Jan.	Feb.	Oct.	Nov.	Dec.	Rainy season
Tra My	70.11	95.85	53.12	52.18	55.47	35.67
Tam Ky	73.56	85.56	47.47	55.01	78.63	36.77

Source: Tra My and Tam Ky meteorological stations, Quang Nam Province

last 30 years in this area, according to meteorological data recorded by Quang Nam's meteorological stations, which also confirms the respondents' concerns about flash floods destroying crops. These obviously have negative implications for agricultural production, in particular, and local daily life in general. Once again, an analysis of the results of meteorological data confirmed the primarily qualitative data obtained from IDIs, KIIs and FGDs – that there has been a high risk of flash floods and great fluctuations of rainfall in the area during rainy seasons, just as respondents stated in interviews.

Dry season rainfall trends

Trends of rainfall amounts within dry seasons from both stations can be approximated using the following figures and table (Figures 2.7 and 2.8, and Table 2.9). According to statistical analysis regarding rainy-season months, the monthly rainfall from June to September was undoubtedly high but quite stable (around 40–80%). Nevertheless, results from other months – March, April and May – were significantly higher than average; the results were 144.6%, 172.1% and 107.6%,

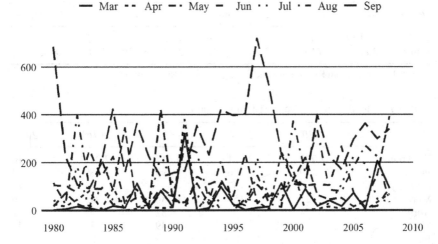

Figure 2.7 Monthly rainfall (mm) in the dry season: Tam Ky station from 1980 to 2008
Source: Tam Ky meteorological station, Quang Nam Province

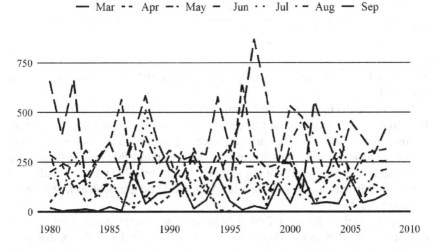

Figure 2.8 Monthly rainfall (mm) in the dry season: Tra My station from 1980 to 2008
Source: Tra My meteorological station, Quang Nam Province

respectively. These numbers suggest huge changes in rainfall in the area from year to year, with particular variance concentrated during these three months.

In other words, rainfall amounts in March, April and May were completely inconsistent throughout these years. There could be extremely dry weather with less than 10 mm/month (as in 1981, 1982, 1984 and 1996), but there could also

Table 2.9 CV values of months in the dry season (CV%)

Station	Mar.	Apr.	May	Jun.	Jul.	Aug.	Sep.	Dry season
Tra My	94.54	76.45	47.51	38.76	64.31	52.91	47.31	18.91
Tam Ky	144.57	172.14	107.60	69.22	82.68	88.59	54.66	28.55

Source: Tra My and Tam Ky meteorological stations, Quang Nam Province

Table 2.10 Monthly rainfall (mm) at Tam Ky station

	2010	2014	2015	2016	2017
Total rainfall	2,705	2,431	2,617	2,213	3,310
January	149	86	93	89	233
February	–	157	–	26	147
March	25	152	36	215	36
April	28	97	32	99	32
May	40	28	49	14	41
June	23	29	48	24	114
July	160	46	206	58	292
August	277	46	33	170	177
September	196	428	101	273	143
October	631	549	879	341	512
November	1,089	770	357	493	1,233
December	87	44	785	412	350

Source: Quang Nam Statistical Yearbook, 2017

be unusually heavy rains, resulting in more than 500 mm/month (as in 1986 and 2000).

In sum, the fluctuations of both temperatures and rainfall data were extremely high from year to year within this period. It means that these two factors were almost unpredictable. Putting these together, we again find that although annual data did show a relative stability, monthly data confirm that year-by-year there were increasing occurrences of hot and dry months or cold and wet months that could result in extreme harm to people's living conditions as well as to their agricultural production.

Reinforcing these findings, Table 2.10 presents more recent meteorological data collected at the Tam Ky station from 2010 to 2017. The data indicate that even during the years considered meteorologically stable, when total rainfall did not change much, fluctuations based on monthly rainfall were consistently high. We should note that 2017 also showed a dramatic increase in both total rainfall and monthly rainfall, said to be in part a result of the La Niña phenomenon that year.

Climate-change estimates for Nam Giang District also show that the mean temperature change in 2050 is projected to increase by 1.9°C (shown in Figure 2.9), which is very close to the scenario of temperature increases up to 2°C that are considered particularly dangerous (UNDP, 2018). According to Dosio et al. (2018), increases of even 1.5°C would result in nearly 700 million people, or

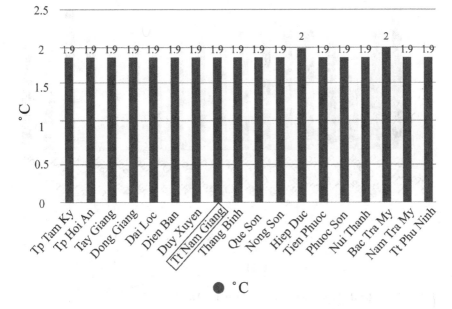

Figure 2.9 Mean projected temperature change in 2050, degrees Celsius per year, RCP 8.5
Source: Viet Nam's Climate Index (UNDP, 2018)

9% of the world's population, being exposed to severe heat waves at least once every 20 years. Moreover, under the scenario of a 2°C warming, the rate would be expected to increase threefold, affecting 2 billion people, or 28.2% of the world's population.

Similarly, the area under study (shown here within Nam Giang District) is projected to be at high risk (70–90%) of rainfall change hazards by 2050, as can be seen in Figure 2.10 (Tp Tam Ky and Bac Tra My locations).

Evidence and consensus: comparing statistical evidence and people's perceptions

To verify evidence of climate change in the study site, comparisons between people's perceptions and statistical evidence of changing factors show a close correspondence between the two. First, statistical data affirmed local perceptions that climate factors have fluctuated significantly over the years.

The second change noticed by respondents was the increase in harsh weather, with hotter summers and colder winters year after year. Women, in particular, noted weather extremes and the damage they could do to agricultural production. As discussed earlier, even though there was no increasing or decreasing trend

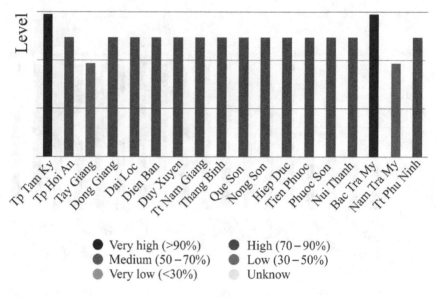

Figure 2.10 Rainfall change hazards in 2050, RCP 8.5

Source: Viet Nam's Climate Change Index (UNDP, 2018)

regarding temperatures from year to year (the annual mean temperatures stayed the same), temperatures have fluctuated significantly from month to month, especially during the summer and winter. As a result, temperatures during summer months have increased year after year, and temperatures during winter months have decreased year after year. The statistical data thus explain and support why local villagers have felt that the weather has become more extreme and, from month to month, so much more unpredictable.

Third, there is a consensus reported among respondents about the increasing frequency of dry spells and flash floods, both of which have had serious negative impacts on people's lives and on agricultural outcomes. Statistical data also confirm such changes as an increasing number of dry spells and flash floods over the last 20–30 years. The results of meteorological data analyses indicate that as a consequence of rainfall and temperatures fluctuating significantly from month to month, the number of hot and dry days potentially leading to dry spells has increased. The number of days with too much rain in a short period of time, causing flash floods, increased as well.

There were a few comments by respondents about the weather that were not supported by meteorological data. First, some respondents stated that the weather has generally become increasingly and noticeably hotter over the years. However, based on statistical meteorological data analysis, annual mean temperatures have not changed. The reason why people had the feeling of hotter weather in general is likely because there are actually some months in which temperatures have increased year after year (same month, different years) and have become significantly higher than in other months of that same year (different months, same year). This wide variation in temperatures may have magnified people's perceptions.

Respondents also stated that there had been less rainfall than in the past. In fact, based on statistical meteorological data analysis, the total annual rainfall has not changed much over these years; however, it has been unevenly and unpredictably distributed over the months. Heavy rains coming in a short period of time lead to more flash floods. This has also resulted in more dry spells because the number of hot days without rain has also increased.

Thus, we find that both qualitative primary data and statistical data from the two meteorological stations confirmed three major changes in climate: *Climate has changed* with unpredictable fluctuations of temperatures and rainfall; the *weather has become more extreme*, with hotter summers and colder winters year after year; and the *risk of flash floods and dry spells* has increased.

Summary

This chapter began with an overview of the Co Tu ethnic minority in the study area, and reasons why women have traditionally held lower status than men in the community. It also discussed the community's spiritual and other cultural beliefs and the importance of their ties to the surrounding forest environment.

The chapter then went on to discuss the impacts of sedentarization programs on the Co Tu community members' lives and their vulnerability to climate change. Interviews indicated that local people were very aware of major weather-related changes and events in recent decades and of the very negative impacts these changes have had on their lives. Specifically, people's perceptions that climate factors – including rainfall, temperature and the timing of seasonal weather – have widely and unpredictably fluctuated are affirmed by the results of the meteorological data analyses presented in the second half of this chapter.

Although local perceptions regarding climate change and meteorological data analysis have agreed on general findings, it was found that the impressions and concerns of Co Tu women and men were slightly different. Women were likely to be more concerned about extreme weather events and the risk of flash floods and long-lasting periods of heavy rain because of their potentially severe impact on agricultural production in general and the rice crop in particular (their most important means of survival). Meanwhile, men were more worried about increasingly hot weather, which they perceived to have a negative impact on living and working conditions. These differences may be connected to their gender-differentiated roles and livelihood tasks, given that women have been the primary agricultural laborers in the community, whereas men's tasks were traditionally more focused on hunting, trapping, cutting down large trees and other responsibilities.

Gender concerns will be discussed in detail in the following chapter, focusing on changes in women's and men's livelihood activities and their roles in the household and community in recent years. These changes came as a consequence of climate threats and other additional compounding factors that have had significant impacts on women's and men's attitudes and behaviors and have resulted in major changes in women's and men's wellbeing in the community, as will be presented in detail in Chapter 3.

Notes

1 Those interested in the history, culture and belief systems of the Co Tu in this region, including but not specifically focused on Ca Dy Commune, are strongly encouraged to read accounts given in Arhem's 2009 publication regarding the French and American wars, as well as the years leading up to and including the 2003–2005 period in which the Katuic Ethnography Project conducted their main field visits. (The Katuic Ethnography Project was a collaboration between the Anthropology Department of Gothenberg University and the Vietnam Museum of Ethnology.) See also Arhem's 2014 study on Katuic cosmology and change in Laos and Viet Nam in the wake of development policies (based on periodic field visits to Viet Nam in 2004, 2006, 2008 and 2009), among other ethnographic studies of the region, including those associated with the Katuic Ethnography Project.

2 A bride price is paid by the groom's family to the bride's family at the time of marriage (it does not go to the bride herself). This is the opposite of a dowry, which is paid by the bride's family to the groom's family (a dowry can also take the form of property, for example, furniture or other assets, given to the new couple from the bride's family). Although in the Co Tu case the bride's family also gives some gifts to the groom's family, the heaviest expenses are paid by the groom's side.

3 See, for example, Thanh Van Mai and Phuc Xuan To (2015, p. 169). "Swidden cultivation has been perceived by most governments in Southeast Asia as a primitive cultivation method that is destructive to the environment (Dove, 1983) and many have tried to ban the practice. However, these efforts have largely failed due to the lack of a comprehensive understanding of this practice. Dove (1983) points to three prominent myths about swidden cultivation that have resulted in an oversimplified understanding of this traditional cultivation method: that the practice is based on primitive communalism; that the practice represents a misuse of the environment; [and] that areas practicing swidden cultivation are isolated from regional, national and international economies."
    www.researchgate.net/publication/330652388_A_systems_thinking_approach_for_achieving_a_better_understanding_of_Swidden_cultivation_in_VietNam) → Hum Ecol (2015) 43:169–178 DOI 10.1007/s10745–015–9730–8

4 In his publication based on fieldwork carried out in 2003–2005, Arhem (2009) notes, "Although the Katu belief system might appear irrational to an outsider, it constitutes precisely the moral basis and motivating force underpinning the local resource-use system. In the past, these beliefs effectively guaranteed a rational form of shifting cultivation that has been environmentally sustainable until very recently. Preliminary readings of satellite images of the Katu landscape from the 1970s onwards confirm this conclusion (Koy et al., 2007)" (p.188).

5 It is bordered by the town of Thanh My in the north, by Phuoc Son District in the south, by Que Lam Commune and Que Phuoc Commune in the east and by Tabhing Commune in the west. Nam Giang District is located in a mountainous area in the west of the province.

6 During one of the first study visits, the research team also experienced an incident similar to the mountain landslides mentioned in this quote. One of the field trips was interrupted because a prolonged period of heavy rain caused a rockslide that prevented any form of transportation from getting through. During FGDs, respondents also mentioned

transport difficulties that had recently been caused by prolonged heavy rains. In another example of extreme weather, during data collection in 2019, the research team experienced abnormally hot and dry weather in the month of March. This was a month that was supposed to have brought in the spring rains, but the rains did not come (it had been nearly two months without rain in the area).

7 Test S is computed by the following formula:

$$S = \sum_{k=1}^{n-1} \sum_{j=k+1}^{n} \text{sgn}(x_j - x_k),$$

where x_j and x_k are the annual values in years j and k, $j > k$, respectively

$$\text{sgn}(x_j - x_k) = \begin{cases} 1 & \text{if } x_j - x_k > 0 \\ 0 & \text{if } x_j - x_k = 0 \\ -1 & \text{if } x_j - x_k < 0 \end{cases}.$$

References

Arhem, N. (2009). *In the sacred forest: Landscape, livelihood and spirit beliefs among the Katu of Vietnam.* SANS Papers in Social Anthropology, 10. Gothenburg: University of Gothenburg.

Arhem, N. (2014). Forests, spirits and high modernist development: A study of cosmology and change among the Katuic peoples in the uplands of Laos and Vietnam. *Uppsala Studies in Cultural Anthropology, 55.*

Arhem, N., & Binh, N. T. T. (2006). Road to Progress? The socio-economic impact of the Ho Chi Minh highway on the indigenous population in the Central Truong Son region of Vietnam. (Presented for WWF Indochina.) www.academia.edu/10639717/Road_to_Progress_The_socio-economic_impact_of_the_Ho_Chi_Minh_Highway_on_the_Indigenous_Population_in_the_Central_Truong_Son_Region_of_Vietnam

Asian Development Bank (ADB). (2010, June). *Commune profile and investment plan: Cady Commune, Nam Giang District, Quang Nam Province.* www.gms-eoc.org/uploads/resources/232/attachment/QN_Cady_english_11Sept.pdf

Bayrak, M. M., Tran, N. T., & Burgers, P. (2013). Restructuring space in the name of development: The socio-cultural impact of the forest land allocation program on the indigenous Co Tu people in Central Vietnam. *Journal of Political Ecology, 20*(1). https://doi.org/10.2458/v20i1.21745

Ca Dy Commune People's Committee. (2013). *Annual socioeconomic reports from 2012–2013.* Vietnamese Version. London: World Bank.

Dosio, A., Mentaschi, L., Fischer, E. M., & Wyser, K. (2018). Extreme heat waves under 1.5 °C and 2 °C global warming. *Environmental Research Letters, 13*(5). https://doi.org/10.1088/1748-9326/aab827

Dove, M. (1983). Theories of swidden agriculture, and the political economy of ignorance. *Agroforestry Systems, 1*(2), 85–99. https://doi.org/10.1007/BF00596351

Epprecht, M., Müller, D., & Minot, N. (2011). How remote are Vietnam's ethnic minorities? An analysis of spatial patterns of poverty and inequality. *The Annals of Regional Science, 46*(2), 349–368. Springer-Verlag (Original work published in 2009). https://doi.org/10.1007/s00168-009-0330-7

General Statistics Office of Vietnam (GSO). (2015). *Results from the survey on the socioeconomic situation of 53 ethnic minority groups 2015 (There are 54 recognized ethnic* groups in Vietnam; the 54th is the Kinh, or ethnic majority group). Washington, DC: World Bank.

Ho, L. T. K., & Umitsu, M. (2011). Micro-landform classification and flood hazard assessment of the Thu Bon alluvial plain, Central Vietnam via an integrated method utilizing remotely sensed data. *Applied Geography, 31*(3), 1082–1093. https://doi.org/10.1016/j.apgeog.2011.01.005

Imai, K., Gaiha, R., & Kang, W. (2007). Poverty, inequality and ethnic minorities in Vietnam. *International Review of Applied Economics, 25*(3), 249–282. BWPI Working Paper 10. https://doi.org/10.1080/02692171.2010.483471

ISPONRE. (2009). *Vietnam assessment report on climate change (VARCC): Institute of strategy and policy on natural resources and environment (ISPONRE)*. https://environmentalmigration.iom.int/viet-nam-assessment-report-climate-change-varcc

Koy, K., Laverty, M., Horning, N., & Sterling, E. (2007). *Improving biodiversity conservation in threatened landscapes of Central Vietnam*. Center for Biodiversity and Conservation, American Museum of Natural History. http://gif.berkeley.edu/documents/Threatened_Landscapes_of_Central_Vietnam.pdf

Luu, H. (2007). *A contribution to Katu ethnography: A highland people of central Vietnam*. SANS Papers in Social Anthropology, 9. Gothenburg: University of Gothenburg.

Oxfam. (2008). *Vietnam: Climate change, adaptation and poor people*. Oxfam. www.preventionweb.net/files/7377_vietnamccadaptationpoverty.pdf

Plant, R. (2002). *Indigenous peoples/ethnic minorities and poverty reduction: Regional report*. Asian Development Bank. https://think-asia.org/handle/11540/2968

Quang Nam Province Statistics Office. (2014). *Quang Nam statistical yearbook 2013*. Hanoi: Thong ke.

Quang Nam Province Statistics Office. (2017). *Quang Nam statistical yearbook 2016*. Hanoi: Thong ke.

Quang Nam Province Statistics Office. (2018). *Quang Nam Province statistical yearbook 2017*. Hanoi: Thong ke.

Salemink, O. (2003). *The ethnography of Vietnam's central highlanders: A historical contextualization, 1850–1900*. London: Routledge.

Tran, M. H. (2009, February 10–12). *Food security and sustainable agriculture in Vietnam*. Country Report. Vietnam. Submitted to the Fourth Session of the Technical Committee of APCAEM, Chiang Rai, Thailand. www.un-csam.org/Activities%20Files/A0902/vn-p.pdf

Tran, T. V. (2009). *The Co Tu in Vietnam*. Saigon: Thong Tan Publishing House.

United Nation Development Program (UNDP). (2018). *Vietnam's climate index*. http://eng.climaterisk.org.vn/

Van Mai, T., & To, P. X. (2015). A systems thinking approach for achieving a better understanding of swidden cultivation in Vietnam. *Human Ecology, 43*(1), 169–178. Springer. https://doi.org/10.1007/s10745-015-9730-8

Vo, T. H. (2012). *Assessment of long-term change of land-use in relation to climate and development conditions: A case study for the coastal district of Quang Nam Province, Vietnam* (Master's thesis), Asian Institute of Technology, Bangkok.

Yue, S., Pilon, P., & Phinney, B. (2003). Canadian stream flow trend detection: Impacts of serial and cross-correlation. *Hydrological Sciences Journal, 48*(1), 51–63. https://doi.org/10.1623/hysj.48.1.51.43478

3   The Co Tu of Ca Dy Commune

Vulnerabilities in the face of rapid change

Impact of regulatory and socioeconomic changes on women's and men's work in Ca Dy Commune

As discussed in the previous chapters, climate change has had a serious negative impact on livelihoods and survival strategies in the Ca Dy Commune region. Other compounding factors have also contributed to ending the self-sufficient form of production and relative isolation of this community, including sudden regulatory and socioeconomic changes. These have, in turn, compelled major changes in gender roles and relations. We will begin with a look back at previous gender patterns and contrast them with the new gender-related roles and responsibilities that have suddenly emerged in recent years, leading to unanticipated challenges to existing gender hierarchies.

Women's and men's work in earlier decades

For generations, the Co Tu of Ca Dy Commune have been fully reliant on natural resources for their livelihoods. Unlike many communities located in other communes nearby, in earlier decades villagers in this area cultivated the so-called long-term rice variety that took at least six months for one crop to mature. The women and men of the community also farmed cassava and corn crops, raised livestock, planted fruit trees and fished in the local river, mostly for their own use. Throughout these years, upland rice had been the most important crop, which 20 years ago could provide enough to make up the villagers' main source of food for the whole year. As emphasized by a 63-year-old man, "We have cultivated rice for generations, only to meet our food demand. Rice is critical for our survival." Another woman respondent affirmed this, stating, "With a poor rice crop, we will face hunger throughout the year."

In general, cultivating upland rice crops involves three major steps: (1) preparing the cultivated land before planting a new crop; (2) seeding and weeding; and (3) harvesting. The Co Tu of this area normally start preparing the land in February after Tet (the Lunar New Year), beginning with cutting and burning trees and branches and clearing the fields. Seeds are usually sowed in this area from April to

May, and weeding is done mainly in July and August. The rice flowers in August, and it can be harvested in September or October.

Clearing forestland and preparing the land for a new crop was traditionally assigned to men while women were responsible for cutting down small trees, weeding and burning to clear the land before cultivation. Within this division of labor, the men's role was critical to crop cultivation since cutting down large trees was seen as requiring a great deal of physical exertion. After being sown, crops were "left to nature's care" along with the weeding and other activities carried out by women. The women's hardest task was then harvesting the crops at the end of this production cycle.

In the past, both women and men engaged in collecting non-timber forest products (NTFPs). Regarding livestock raising, however, these activities have traditionally been assigned to women, a girl child, or an elder; when Co Tu men took part in raising livestock, they would consider it as helping their wives, not as their responsibility.

Regarding men's work in earlier decades, back in the 1980s, Co Tu men were well known for being good hunters. They used to consider hunting their main, and by far their favorite, livelihood activity. It was socially designated strictly as a male task. In those years, both hunting and the cultivation of rice crops started in February or during the spring season of each year. For this reason, women took care of farming activities at this time while men focused on hunting.

When the hunting season began, men formed groups to go into the forest together, where they could spend days or weeks hunting. It was not only a livelihood activity, it was also and more importantly a passion of the Co Tu men. It represented a symbol of masculinity. More than just meeting food needs, it was also viewed as a means of protecting their families and community from animal attacks.

Hunting is thus a reflection of one's masculinity as a protector and provider in this context (Luu, 2007). During hunting season, the Co Tu people could have a good number of meals relying on the meat of the animals they caught. Hunted or trapped animals that were large – such as deer, bear, or wild pig – were often shared with the entire village. Small animals, such as porcupines or rabbits, would be used for the hunter and his family's daily meals. The meat was also dried or smoked to save for special occasions. Respondents referred to this time as the "old happy days" when people were not exposed to the cash economy and enjoyed sharing with each other, ensuring close and interdependent relationships (and, one might add, these were also the days of closeness after earlier years of war and separation had come to an end).

All of this started to change in the 1990s. In the next section, the key regulatory changes from the 1990s onward will be summarized, and their impact on men's and women's livelihood activities will be introduced. This will be followed by a discussion of other sociocultural and economic changes that have had a significant impact on gender relations, and it will be argued that all three driving forces – climate change, regulatory change, and socioeconomic change – have had a major impact on transforming gender roles, relations and hierarchies in the Co Tu community of Ca Dy Commune.

Regulatory changes in land use and forest protection

Until the 1990s, the land and forest resources in Viet Nam were managed by the state. As a result of the Doi Moi reforms of 1986, introduced by the Government of Viet Nam to transform a centrally planned economy into a socialist-oriented market economy, the Land Law of 1993 began to be enforced in all rural areas of the country. With the implementation of the Land Law, for the first time, individual households—including those in Ca Dy Commune—were allocated their own cropland. This regulatory reform brought along many changes in the country's land use management system.

By the 1990s, new forest protection policies were also put in place. In terms of biodiversity, Viet Nam has been identified as one of the most biodiverse countries in Southeast Asia and the 16th most biologically diverse country in the world (Vuong, 2008), with a historically substantial forest cover. However, the amount of forest cover has decreased over the years, from 43% in 1943 to 30% in 1985 to 27.8% in 1990. In 2006, as a result of the reforestation program known as "Five Million Hectares Reforestation" (661 Program), the forest cover of the country increased and was said to account for 38% of the total land area (Vuong, 2008). In spite of this, the natural forests in Viet Nam are still suffering from deforestation and forest deterioration.

Recognizing the problem, the government initiated a series of regulatory changes, policies and programs to increase the forest cover and protect forest resources. A number of new laws were passed in order to protect both the forests and wild animals living in them, which were perceived as being under threat. These new laws included such provisions as a ban on hunting and logging and restrictions on trapping and the collection of NTFPs in forest areas. The new laws thus had major impacts on local forest-dependent communities, including the men and women of Ca Dy Commune. Whether the new laws could succeed in achieving the stated goal of forest protection by being applied without adjustment to local social and environmental conditions is a key question, as will be discussed in this and the chapters that follow.

In the Co Tu case, given the fact that the community is forest-based and forest-dependent, the changes related to forest management and protection have had strong and direct impacts on the community. As shown in Table 3.1, which summarizes the most important regulatory changes that have been introduced since 1986, the Government of Viet Nam has been focusing in particular on the development of forest resources. In 1991, the Law on Forest Protection and Development stipulated that forest resources could be allocated to organizations and individuals for management, protection and commercial purposes. It also established the legal basis for setting up management boards for forest protection and special forest uses.

Table 3.1 Regulatory changes that have been introduced since 1986

Year	Policy/Law	Objective/Rationale
Late 1986	Doi Moi reforms (Renovation Policy) were introduced by the Government of Viet Nam.	To transform a centrally planned economy to a socialist-oriented market economy.
1991	The Law on Forest Protection and Development was enforced.	To improve forest protection and management. (Hunting and swidden cultivation on forestland are forbidden.)
1992	National Program 327 was introduced.	To enable individual households to access annual contracts for forest protection, restoration and regeneration.
1993	The Land Law of 1993 was implemented.	To address land rights, for the first time, by allocating cropland to individual households for their own production and giving farmers the right (the so-called Red Books) to inherit, mortgage, transfer, exchange and lease land.
1998	The Forest Land Allocation (FLA) program, one of the subcomponents of the Land Law, was implemented.	To allocate forestland to organizations and individuals. To meet the objectives of the Five Million Hectares Reforestation program (661 program). The FLA program can be seen as one of the strongest sedentarization efforts of the government to date (Plant, 2002; Vuong, 2008).

Source: Data collected from fieldwork

In 1992, National Program 327 entitled individual households to receive annual contracts for forest protection, restoration and regeneration. Households could also receive cultivable land for agroforestry or agricultural purposes. This program was designed and implemented along with the re-greening of bare lands and denuded hills.

As noted earlier, the Land Law of 1993 was established to give farmers the right (the so-called Red Books) to inherit, mortgage, transfer, exchange and lease land. In accordance with the Land Law, Decree 02/CP was enforced and enabled the State to allocate forestland to organizations, households and individuals for long-term use (50 years).

This combination of prohibiting forest exploitation for the purpose of cultivation and the cultivated land allocation as a result of the Land Law of 1993 resulted in the strict forbiddance of swidden (shifting) cultivation. Upland households were required to move from swidden cultivation on forestland to fixed cultivation on their allocated land. These regulatory changes were expected to have important and positive impacts regarding both *forest protection* and *agricultural incomes*, as swidden systems were considered by many policymakers to be less productive (less "modern") and more environmentally harmful.[1] However, the forest-based community in this part of Viet Nam was actually negatively affected because this type of fixed cultivation system was blamed for a decrease in the land available to them for agricultural production. Moreover, they found that fixed cultivation on the same slopes over time could lead to soil degradation because of overuse and soil erosion.

Furthermore, according to Bayrak et al. (2013), from the perspective of indigenous (ethnic minority) communities in Viet Nam, the lands allocated to indigenous households were often fragmented and located far from their houses. More than 80% of respondents interviewed for this study reported that they now own more than three pieces of cultivated land, which are generally small and scattered far from each other and from their homes.

It was also confirmed in the course of interviews with some officials for this study that Kinh (ethnic majority) migrants and other influential ethnic groups tended to benefit more from the land allocation policy of the Land Law of 1993 than was true of the ethnic minorities living in these more remote areas. The Kinh are believed to have more political power and better access to social networks, according to interviews with key informants. Co Tu elders also noted in interviews for this study that, in their case, the land allocated to ethnic minority households is often infertile and therefore not suitable for tree planting or many other types of cultivation over a long period of time, and much of the land that has been available for allocation is too far from people's settlements and villages. The following quotes from in-depth interviews (IDIs) summarized their new circumstances:

> *After Doi Moi, the government distributed land to us. We were provided with the Red Books. We were not allowed to cut old forest for cultivation. We started cultivating only on our allocated lands. However, these cultivated lands are generally very bad, lying far away and having low rice yields.*
> *(IDI with the village patriarch of Pa Cang Village)*

> *Our lands are far apart and difficult to cultivate, with bad soil quality. In the past, we did not cultivate on those lands. We preferred cutting the old forestland because it was rich, so it provided high yields.*
> *(IDI with a 74-year-old man, a village elder of Pa Bang Village)*

A further development came with the 1998 Forest Land Allocation (FLA) program, which was designed with the main purpose of restoring the forest cover to at least 43% by 2010 in order to meet the objectives of the Five Million Hectares Reforestation program (Vuong, 2008). The FLA program can be seen as one of the strongest sedentarization efforts of the government to date (Vuong, 2008; Plant, 2002). The rationale of the program was that if villagers had formal rights to forestland, they would take greater care in forest protection and management. Villagers who received forestland were paid for protecting the forests, and they were given subsidies for tree planting. Furthermore, the legislation on benefit-sharing related to forests (Decision No. 178, November 12, 2001) allowed individuals and households to receive two-thirds of the total value of harvested products, and one-third of the share would go to the commune or other government entities.

However, according to an interview with a forestry expert who worked with forest-based ethnic minorities for many years, local ethnic minority communities did not generally benefit from these policies as they were not able to protect the forest from illegal logging. Moreover, we can see that the FLA program did not work effectively in this study area. First, the economic profit from this activity was not considered to be large enough to make villagers interested in participating. According to the Decision No. 661/QĐ-TTg, a household was given financial support of less than 1 million VND (71.4 USD, using exchange rates at that time) for six years of planting and nurturing each hectare of forestland for purposes of reforestation. According to data collected from IDIs and FGDs, this amount of money was not enough to plant, nurture, and take care of a hectare of reforested area within six years, particularly in view of the fact that planting the forest takes at least ten to fifteen years before big trees can be gathered and utilized. Given that the Co Tu community was very poor, interviewees noted that it was impossible for people to survive up to the point when they could exploit the forest they had planted, even if they had known how to sell the forest products in the relevant markets (which they did not at that time).

Apart from this initial financial support, households involved in the program were also allowed to sell and make a profit from the wood they planted under two conditions: (1) the household needed to inform the forest protection agency in order to get certification that the wood was legal and allowed to be sold in the market, and (2) the household was responsible for replanting the forestland within two years. This constituted another reason the Co Tu community hesitated to join the program: They thought that selling big trees would be out of their capacity, and, as noted previously, they were not familiar with the market, given that at the time the program was launched the Co Tu population was still very isolated. In addition, the new approaches to forest plantation and management were entirely unfamiliar to the Co Tu people, who had been used to swidden cultivation for generations. Therefore, they were not confident that they could practice the new planting method, even if they were given trainings by the government:

> *No, we did not register for the reforestation program. We had never done that before. We did not know if we could do it. Besides, the money paid for the activity was very low. It was not enough to cover the costs of the plantation. We also needed to eat, to support our children. That's why most of the people in my village were not interested in the program.*
>
> *(IDI with a 54-year-old woman)*

The findings from Bayrak et al. (2013) indicated that the FLA program posed a number of problems for ethnic minority communities and thus was not successful in the communities they studied. These findings were confirmed by respondents during interviews conducted for this research. The following quotes illustrate the local Co Tu thinking about the FLA program:

> *No, we did not participate in the reforestation program because the forest-land allocated under the program was normally very far from our homes. We cannot travel regularly to this forestland to take care of it.*
>
> *(FGD, 56-year-old man)*

> *Profit from forest reforestation is very low, so we don't want to participate in the program. It takes too much work, and it does not provide us with enough to live.*
>
> *(IDI with a 51-year-old man)*

Moreover, as one woman put it, men were reluctant to become hired laborers on forestland:

> *We don't have enough laborers to be part of the reforestation program. I am busy with our farm. My husband just does not like working for other people because it is uncomfortable for him. Also, just like many other Co Tu men, he does not like to be seen as a hired laborer working for a boss.*
>
> *(FGD, 42-year-old woman)*

In other words, men were used to their traditional roles as local residents and careful users of the forests, not as wage laborers having to take orders from an employer for the commercial exploitation of the forests. This is an important point that will be taken up again in the following chapters.

In sum, the regulatory changes discussed earlier have had very serious impacts on both Co Tu women and men. First, due to the prohibition of using forestland for cultivation, the local population now faces a lack of cultivated land with good soil quality, and they have had to shift from swidden cultivation on natural forest-land to cultivating on their own allocated pieces of land. Second, food security has been negatively affected by the reduction in soil quality, together with the ban on hunting and logging and restrictions on trapping and collecting NTFPs in forest areas. Third, social and cultural customs and norms have been accordingly affected by the loss of or changes made in traditional livelihood activities. Among other developments, the village patriarchs started playing a less important role in their villages, and the local people's cultural lives have been changed due to the loss of indigenous knowledge and traditional forest management practices that worked well in the context of a local community producing to meet its own needs.

It is worth reemphasizing here that even less than ten years before the regulatory changes were enforced, the community was still relatively isolated and self-sufficient. Solidarity was one of the Co Tu's strengths and core cultural values as it served to protect the forest-based community from animal and other threats and respond effectively to natural disasters. Their spiritual lives were centered on the surrounding natural environment and on community activities, and they were strong as a group; as an iconic example of this, men would go hunting as a group, and the large animals caught during the hunts were shared and celebrated with the whole village. When the government's programs and new regulations approached them as individual households, they were said to have broken their sense of con-nectedness, group strength and cultural values. One example of this was the development that, as individual households, the communities were no longer able to protect their allocated forestlands from illegal logging, a change noted by the forestry expert. The extent of social disruption caused by these changes cannot be overemphasized, particularly given that the transitions were required to take place in such a very short period of time.

Gender and livelihood patterns

From a gender perspective, since the traditional designated roles highlighted men as hunters and women as key swidden farmers, regulatory changes – banning shifting cultivation, hunting, and logging (cutting large trees) and with restrictions on trapping and collecting NTFPs in forest areas – have affected Co Tu women's and men's lives in very disruptive and difficult ways. Men lost their key livelihood activities, and women struggled to continue farming their small areas of allocated land.

This shift had significant impacts on the gender division of labor among Co Tu families, which led to major changes in women's and men's roles, relationships and hierarchies within their families and community. On one hand, men were no longer able to contribute to their households and the community through their hunting and related forest-based activities, and they were also no longer criti-cally needed for livelihood activities that required significant physical strength. Women, on the other hand, could no longer select good pieces of land for their farming activities, which negatively impacted their agricultural production. The regulatory changes, therefore, have made both women and men face food and livelihood insecurity; moreover, the changes have not only limited their ability to fulfill their expected roles, they have also disrupted the family's traditional struc-ture and the cooperative and complementary relationships between husbands and wives with regard to the family's livelihood activities.

As discussed previously, when swidden farming was freely practiced in for-ested areas, although women were still the key laborers, men also played an important role in cutting trees in forest areas that had not been recently cultivated and helping clear the land to prepare for the new crop. Such practices made men more centrally engaged in this activity alongside the women. In addition, since in earlier years the local swidden practices involved leaving the cultivated land fallow for many years and protecting certain parts of the forest so that it could cither remain pristine or regrow, the local people could circulate far away from their homes and build small temporary houses near the fields to stay during the period of land preparation, weeding and harvesting. In this context, men engaged more with agricultural production and often stayed in the temporary shelters, par-ticularly during the time of land preparation and harvesting because it took time to travel home, and threats from wild animals were a significant concern when traveling through fields, clearings and thick forests. In contrast, when local households were given fixed plots of land to cultivate, the men's engagement was no longer considered important since preparing the land no longer involved cutting large trees and the practice of building temporary houses and staying near very distant fields was no longer required. Farming activity thus came to be progressively transferred into women's hands since men's contributions were no longer seen as important to agricultural activities.

In this way, women's work and their time spent on farming activities have increased not only because of the reduction in men's contributions but also because their fields are now located in fixed areas. These areas are not so far away that they have to walk for days as they did in the past, but they are still a significant distance from the women's homes. More than 70% of respondents surveyed for this study noted that it could take them one to one-and-a-half hours to walk to their fields, and the women do not stay in the fields but walk home every day. In addition, around 13% of respondents interviewed had to use a boat to carry the rice home because their fields were now located across the river, away from their houses.

Women harvest little by little, gathering the grains of paddy (rice still in the husk) with their hands and fingers and putting the paddy in small bags, and later carry the paddy home in big bags called gùi. The harvest thus takes a great deal of time to complete, from ten to fifteen days to a month, because it is done manually, mostly by the Co Tu women themselves. In some cases, the harvest could take even longer:

> Our cultivated land is over there, on the other side of the river. It takes me more than one hour to walk there because we do not have a boat or anything else to cross the river more quickly. As a result, I would need one or more months to harvest our crops. For this reason, if a flood comes earlier than expected, we could lose part of or even the entire crop.
>
> *(IDI with a 44-year-old woman)*

Such impacts could incur great losses, leading to a high degree of food insecurity. For this reason, the majority of respondents agreed that although climate change was the most important cause of the reduction in rice yields, the second most important factor was regulatory change restricting where they could cultivate; both contributed to greatly increasing the community's food insecurity.

In addition, based on data-collection results, particularly from the 2011–2012 and 2014 visits, there was no indication that Co Tu men replaced their old activities, which were no longer available or relevant, with new types of livelihood activities during the years following the enforcement of the new laws and programs. Since hunting was now strictly forbidden, trapping was restricted and cutting big trees in old forests to clear land was no longer allowed, the Co Tu men felt they were no longer needed, and therefore they did not want to work in the fields as much or as regularly as the women did (it is also likely that they did not want to take on what was viewed as "women's work"). The impact this had on gender roles and responsibilities will be explained in greater detail in the following chapter.

First, though, it will be important to discuss one additional factor that, together with the changing climate and regulatory framework, pushed the community away from self-sufficient production and into greater contact with the mainstream economy and society of Viet Nam. This also contributed to the sudden and very disruptive changes that have affected the entire community, with both positive and negative outcomes.

Socioeconomic and cultural change

In addition to climate and regulatory change, a third driving factor – socioeconomic and cultural change – has had a major impact on both women and men in the community and on younger generations as well. We will first outline some of the most important social changes that have come to, and been imposed on, the Co Tu of Ca Dy Commune. Then, in the following chapter, we will detail how the combination of all three of these changes – climate-related, regulatory and socioeconomic – have led to significantly greater vulnerabilities in the Co Tu community in recent years, which, in turn, have had a major impact on gender roles and hierarchies as well as other important consequences.

In the past, the Co Tu community in this area lived in small groups or so-called "villages," and Ca Dy Commune has nine villages. These nine groups are located far away from each other, and it would take half a day or even a full day to go from one village to another because in some locations walking is the only way to travel in this forested, mountainous area.

The first and most critical event that influenced women and men socially, economically and culturally was the construction of National Route 14B (or National Highway 14B), which began construction in 2004 and was completed in 2005. National Route 14B is one of the most important national highways in Viet Nam because it connects the northern central provinces to the southern Central Highland provinces and the eastern provinces of the (more general) southeastern region. National Route 14B is also connected with Lao PDR's National Highway 13 as part of the Asian road network. The completion of Route 14B ended the isolation of Ca Dy Commune, leading to a series of socioeconomic and cultural changes; it has brought the Co Tu community closer to the outside world and specifically to the Kinh majority.

Another very significant social change tied to the completion of National Route 14B is that from 2008 to 2012 the commune was removed from the list of "especially poor communes" because it was expected that the national highway would bring in new benefits and more prosperity to the area (in this case, the improved infrastructure did add points to the area based on the system of evaluating poverty). However, because the commune's actual socioeconomic conditions did not improve as much as had been expected, in 2012 it was put back on the list. Nonetheless, the years in which the commune was classified as "non-poor" came as a shock to the local community for many reasons.

Being on the list of especially poor communes has been very important to the local people because it qualifies the commune for a government subsidy program that provides (1) support for children in the form of free tuition, subsidies for food and an additional monetary allowance for rated-poor households; (2) 1 million VND (50 USD) for rated-poor households to improve housing infrastructure (primarily sanitation); (3) support for the commune in improving its means of communication throughout the commune; and (4) support for healthcare services. However, when it was removed from this list, the government subsidies, including for children's education, food, money allowances and rice grain given during extreme weather events, were suddenly eliminated.

Even though this was later reversed, the loss of the commune's listing as especially poor has had serious consequences. After they lost the designation, the Co Tu population – which was in fact a very poor population with little experience of the world outside the commune – suddenly faced serious financial difficulties as they now, for the first time, had to pay for children's tuition and health care and cover the losses of other subsidies in a context in which little cash had ever been needed or used. This would have a particularly strong impact on women, who felt they were responsible for meeting their families' basic needs, particularly regarding their children's food, health and educational needs.

These new and significant financial difficulties, together with increased interactions between the Co Tu and other communities as a result of the highway, pushed the Co Tu community in this area into the market economy due primarily to their sudden need for cash. This caused substantial changes in both family activities and relationships and changes in economic activities in the community as a whole.

These changes have had important gender implications because of the highly gender-differentiated livelihood activities of the Co Tu women and men. To the Co Tu community at this time, *small-scale trading* was a relatively new livelihood activity that was introduced and developed by the Kinh majority, who began migrating in small numbers to this area after 1975 (i.e., since the war ended) – and it was one that Co Tu women, not men, took up as a means of survival.

New needs and uses for cash in a previously self-sufficient economy

Along with the construction of National Route 14B, a series of additional social changes were imposed on the commune. First, due to land acquisition for road construction, about 20% of the households in the commune lost part of their cultivated land as it was taken for road construction or hydropower plants. Most of this land had been used for wet-rice cultivation, which did not play as important a role in the agricultural production activities of this area compared to upland cultivation. Second, households whose lands were required for land acquisition were paid a sum of money. Reimbursed money varied from 5 million VND (316 USD, using exchange rates at that time) to 30 million (about 1,895 USD), depending on the household. For most households, this was the very first time they had encountered such a large amount of money. Discussions with the village and commune leaders revealed that nearly 70% of the households that were reimbursed used a large part of the funds to buy televisions, radios, motorbikes, mobile phones and other consumer items. Only around 30% of these households invested a significant part of the reimbursed money toward income-generating activities, such as buying cows or buffalos consistent with their agricultural activities.

This infusion of cash thus contributed to the introduction of consumer goods into the community, as the following quotes suggest:

> *With the reimbursement money, we bought a motorbike.*
> *(IDI with a 47-year-old man)*

> *I wanted to buy a buffalo, but after having paid off a debt and buying a televi-sion, we ran out of that reimbursed money.*
> *(IDI with a 43-year-old woman)*

> *We used the money to buy a pig and some chickens. With the rest, my husband bought a radio and a mobile phone for my son.*
> *(IDI with a 54-year-old woman)*

Aside from the sudden need for and appearance of cash, the completion of National Route 14B increased the interaction between the Co Tu and Kinh communities by facilitating travel to and from the commune, and as a result the Co Tu community was no longer as isolated as it had been in earlier years. National Route 14B facili-tated the development of trading between local communities and outsiders since it is the main route between the north and south, and substantial numbers of trucks and cars pass by every day. To meet the needs of people traveling through the area, new small vendors, hostels and other services increased. This has also contributed to changing the commune's character.

As noted earlier, after the completion of National Route 14B, the Co Tu peo-ple of Ca Dy Commune started learning more about the pull of consumer soci-ety, including such modern commodities as mobile phones and motorcycles, and entertainment mediums, such as video games and karaoke. Since then, the local community's needs have increased beyond food and other requirements as traditionally and locally defined; commercial society, in contrast, would now draw them toward ever-changing and ever-increasing needs, which in turn have required increasing amounts of money to purchase items that are now seen as both desirable and needed.

As a result, the Co Tu people – and men in particular – have started to become interested in acquiring new items, and expenditures changed rapidly. At the same time, due to the change in being designated a "poor commune" to a "non-poor commune," the local community lost government subsidies. The combination of these socioeconomic changes increased the need and desire for cash and what cash can buy, with striking results, as noted in this key informant interview (KII):

> *You know, some men in our village started learning about video games and karaoke. Since the road was finished, these services have been opened in our area to serve truck drivers passing by and staying overnight.*
> *(KII with an officer working for Women's Union of Ca Dy Commune)*

As a result of these rapid changes, respondents commonly observed that young men have developed a habit of gathering at coffee shops and internet shops during daytime hours, and in recent years many of those interviewed (interestingly, these tended to be village elders and women respondents) complained that young unmarried men have become "lazy" and are not concerned with contributing to their families in any way. Women objected to the excessive idleness of both young and middle-aged men in the commune as its character started to change. ("Since the road was completed, men in our village have engaged in many bad habits, such as karaoke and games," according to an IDI with a 57-year-old woman, commenting on men's long hours of idleness.) Usually, the observation was that middle-aged men would waste a great deal of time drinking, and younger men would waste time playing games.

It is important to remember that the Co Tu had virtually no experience in handling large amounts of money until the road construction was started, according to interviewees. Even those who used the money to invest in livestock raising were not very successful because of a lack of animal husbandry knowledge, poor facilities and severe weather conditions – in fact, a large segment of the livestock, including cows or buffalo, did not survive the extreme cold weather during winter months, given that the livestock facilities were usually in very poor conditions and climate extremes were becoming more severe over time. As a result, the money given in compensation for the acquisition of land did not make a significant contribution to developing sustainable livelihood activities but instead resulted in integrating the Co Tu into the market economy in a way that contributed to a lifestyle some reported they could not afford. As one respondent noted:

> *Life has gotten tougher as we now need more money than we did in the past. Children need to go to school and to have as high a level of education as they can because they don't want to be farmers like their parents. We have to pay for their tuition fees. We need to buy them nice clothes, unlike before when we made our clothes ourselves. Apart from that, they ask for a lot of things, such as mobile phones and motorcycles, like their friends have in lowland areas.*
>
> *(IDI with a 43-year-old woman)*

According to the older respondents, until the 1980s life was relatively easy, simple and relaxed. There was little interaction with the outside world, and Co Tu people were notably satisfied with their lives. Although they did not have a cash income and barely had any savings, the elders interviewed in recent years still felt that life was easier at that time. They could produce enough for their basic food consumption, and if they lacked food they could borrow from neighbors or relatives. Before 2008, they said there was little need for cash because the commune's subsidies covered children's tuition fees and health care, as well as much of the money and rice needed for weddings, funerals and other special occasions.

Community members, therefore, were not under pressure to earn money in the same way as after the commune was "opened up" to the outside world. Up to this time the buying and selling of goods remained extremely rare in the Co Tu community because they preferred giving or borrowing over buying and selling. They said that most of the time they did not even have to return borrowed items, knowing that they would take turns helping each other. At that time, the idea of selling products to other members in the community was "awkward," according to older respondents:

> *Many years ago, we were poorer than we are now, as we never had money. Still, we treated each other well. Relationships between neighbors were so good that we never sold to or bought from each other. Back then, we shared when we went hungry. We borrowed from our neighbors without having to return what we borrowed. We took turns helping each other.*
>
> *(IDI with a village elder of Ngoi Village)*

> *I thought it was awkward selling to our neighbors. However, since the Kinh moved here, it has become more common. We sell our products to them and buy from them when we need to.*
>
> *(IDI with a 57-year-old man)*

As discussed previously, government subsidies returned when the commune was put back on the list of especially poor communes. However, the exposure to the outside world and the perceived need to "develop" in order to survive in the new economy and society has put more pressure on the local community to find ways to get cash to meet new purchasing needs. These pressures come on top of the major stresses already imposed on the community by increasingly unpredictable climate conditions and the sudden regulatory changes, leading to major new challenges for women and men, both in older and younger generations, throughout the community.

Gender, age and assessments of socioeconomic change

Interestingly, assessments of gains and losses in recent decades have tended to differ according to gender and, to a lesser extent, age. During the interviews, when asked whether life had gotten better or worse over the last 20–30 years, there were two main types of responses. One set of respondents answered that they thought life had become worse because money requirements and prices had increased a great deal and farming had become more difficult with more severe weather and degraded soil quality. Another set of respondents – mostly young men under 30 years old – thought the quality of life had improved and argued that new technologies, such as motorbikes and mobile phones, had made life easier and that items such as televisions and karaoke machines had made life more enjoyable.

Survey results, conducted when the commune had been taken off the especially poor commune list and consumer goods had started coming in, supported the respondents' ideas expressed in interviews: There were more male than female respondents who thought "life is getting better" (30% of male respondents vs. 18% of female respondents, overall), and once again this was found to be especially true for male respondents who were under 30 years old, as noted in the following IDI:

> *Of course, life has gotten better. We have motorbikes, so traveling to the low-land areas becomes much easier. We have a lot of things that we did not have in the past, such as mobile phones and televisions.*
>
> *(IDI with a 25-year-old man)*

On the other side, 80% of respondents who said "life is getting worse" were female respondents over 30 years old. As discussed in interviews:

> *I think we have to work harder than we did in the past because there have been a lot of new expenses, such as electricity and the need to buy rice.*
>
> *(IDI with a 55-year-old woman)*

> *I have been worried all the time about earning money to cover new expenses, including tuition fees, rice for deficiency days, and many other needs.*
>
> *(IDI with a 44-year-old woman)*

The survey also found that middle-aged men tended to be divided or ambivalent about these new developments, whereas younger men were excited about the changes. In contrast, women felt more burdened by the changes and were particularly worried about meeting expenses so that their families could survive.

Summary

Over the past few decades, the Co Tu community in Ca Dy Commune has been through a turbulent period with a great deal of regulatory and social change that, on top of the serious problems they face due to climate change, have had significant impacts on the local community. This period started with the prohibition of swidden cultivation due to regulatory changes regarding land use and forest protection in 1991, followed by additional laws and programs. Swidden farming was a type of cultivation that was associated with an entire traditional lifestyle and social and cultural system. The disappearance of swidden cultivation forced women to change their farming practices and moreover led to a decrease in men's participation in farming because their role had been centered around cutting big trees to facilitate the women's and men's traditional farming practices.

The second most important change was the hunting ban, originally designed to protect wild animals in forested areas. This removed the most critical livelihood activity of the Co Tu men, which was considered a symbol of masculinity to the Co Tu people. (Observers also noted that the new regulations did not necessarily protect the animals in forested areas, at least initially, as outsiders took advantage of illegal hunting and logging opportunities for cash purposes; these problems emerged as the villagers' earlier shared system of access to forest resources came to an end.) As a consequence, regulatory changes have had important gender implications as they affected both Co Tu women and men in significant but very different ways, with middle-aged and young men becoming more disengaged from livelihood and community activities and women and girls having to take on new roles and burdens to ensure their family's survival.

In addition, a wide range of social changes began with the construction of National Route 14B in 2004–2005 and later with the loss of government subsidies caused by the community's loss of status as an "especially poor commune" (from 2008 to 2012). These changes have resulted in increased interactions between the Co Tu and the Kinh communities by facilitating travel to and from Ca Dy Commune, ending the commune's isolation, and by greatly increasing the villagers' financial pressures and cash-based requirements for a population that rarely used much money or relied on cash-based incomes in the past.

Climate, regulatory and social changes have thus all combined to affect Co Tu villagers in terms of both rapidly reducing food security for families and the community as a whole and forcing changes in their livelihood systems, practices, values and beliefs. The following chapter will look at the shifting relations between husbands and wives tied to changes in their roles and status in the family and in the community. There have also been strong effects on young women and young men as they face these new pressures and difficulties – and indeed, for some, new opportunities as well.

Note

1   Interestingly, a study conducted in four forest-dependent villages in Lao PDR found that incomes derived from land under swidden cultivation are often underestimated (Van Der Meer Simo et al., 2019). The study was undertaken in order to understand why rural Lao farmers continue to prefer swidden techniques to practices policymakers and practitioners generally believe to be more profitable and environmentally protective. It was found that swidden cultivation actually accounted for 75% of the mean annual household livelihood incomes of the Lao cultivators, and they were reluctant to give up these techniques because they found them to be more productive. Moreover, as noted previously, there are questions about which approach is more environmentally sound (this depends on many factors); however, the Lao farmers presumably found these swidden techniques to be sustainable in their social and environmental context because if they found negative environmental impacts, they might be more open to change.

References

Bayrak, M. M., Tran, N. T., & Burgers, P. (2013). Restructuring space in the name of development: The socio-cultural impact of the forest land allocation program on the indigenous Co Tu people in Central Vietnam. *Journal of Political Ecology, 20*(1). https://doi.org/10.2458/v20i1.21745

Luu, H. (2007). *A contribution to Katu ethnography: A highland people of central Vietnam.* SANS Papers in Social Anthropology, 9. Gothenburg: University of Gothenburg.

Plant, R. (2002). *Indigenous peoples/ethnic minorities and poverty reduction: Regional report.* Asian Development Bank. https://think-asia.org/handle/11540/2968

Van Der Meer Simo, A., Kanowski, P., & Barney, K. (2019). Revealing environmental income in rural livelihoods: Evidence from four villages in Lao PDR. *Forests, Trees and Livelihoods*, *28*(1), 16–33. https://doi.org/10.1080/14728028.2018.1552540

Vuong, X. T. (2008). Forest land allocation in mountainous areas of Vietnam: An anthropological view. In S. Robertson & T. H. Nghi (Eds.), *Proceedings of the forest land allocation forum on 28 May 2008* (pp. 45–55). Hanoi: Organized by Tropenbos International Vietnam in Cooperation with the Forest Protection Department with support of MARD.

4 The struggle for food and livelihood security

Changing livelihoods, gender roles and gender hierarchies

Collective impacts of a changing climate and society

We have argued in the preceding chapters that, in many cases, the impacts of climate change should not be considered in isolation; in such cases, they need to be put in the context of other important changes that are taking place. This has been true as well regarding the present study, where women's and men's roles, voices and hierarchies have been affected not only by physical changes in their environment but also by changes in their institutional and social contexts (Chaudhry & Ruysschaert, 2007; World Bank, 2010).

In this chapter, we will examine men's and women's responses to rapid increases in food and livelihood insecurity, with women impacted more by *food insecurity* and men (and women to a lesser extent) by *livelihood insecurity*. We will also examine how relations between wives and husbands have changed and, more generally, how gender roles and hierarchies have changed, in the wake of the women's and men's very different responses to sudden and difficult food and livelihood-related threats.

We will be concerned primarily with the years following the changes that forced the community to move away from self-sufficiency and toward the country's economic and social mainstream. The discussion will focus initially on data obtained during the 2011–2012 and 2014 field visits to the area; more recent – including some unexpected – developments, drawing from interviews held in 2019, will conclude this chapter and serve as a basis for further consideration of changing gender roles and hierarchies in the aftermath of major climate, regulatory and social changes in the next and final chapter (Chapter 5).

The following timeline (Figure 4.1) identifies some of the major turning points for the community from the 1990s onward, with changes compounding especially in the last two to three decades (particularly since 1998, with the enforcement of the Forest Land Allocation [FLA] program and other regulatory policies, along with increasingly significant impacts of climate change and socioeconomic and cultural changes over time).

As noted previously, agriculture is the critical source of livelihood for 97% of the Co Tu population. They have relied on subsistence agricultural production and forest resources since their ancestors' time. However, in recent years, agricultural production and livelihood patterns have been undercut and forced to change. We

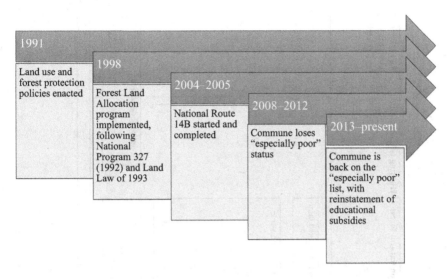

Figure 4.1 Recent turning points impacting the Co Tu community of Ca Dy Commune

will therefore begin with a focus on growing food and livelihood insecurity and the resulting impacts on gender roles, relations and hierarchies.

Impacts on food insecurity

Food security has been under threat particularly because of the decrease in rice production caused by both climate change and regulatory changes. Those interviewed believed climate change – specifically, increasingly prolonged hot and dry weather, intense dry spells, prolonged periods of heavy rains, and flash floods – was the main cause of crop losses. In addition, regulatory changes – which have led to a reduction in cultivated land and a consequent degradation of the soil – were found to be the second most important reason, according to respondents, that has caused falling rice yields and reduced amounts of agricultural production in general.

Figure 4.2 represents the combined impacts of regulatory and climate changes on the food security of the Co Tu community. As a consequence of these changes, most of the respondents were concerned that although they were able to produce enough (or close to enough) rice for a year's worth of household consumption back in the 1980s, they are no longer able to do so, as seen in the following quotes from in-depth interviews (IDIs):

> *In the old days, twenty or thirty years ago, we grew enough rice to meet the whole year's rice demand. Things have changed. The rice that we produce now is just enough for our household's use for three to five months.*
>
> (IDI with a 73-year-old man, village patriarch)

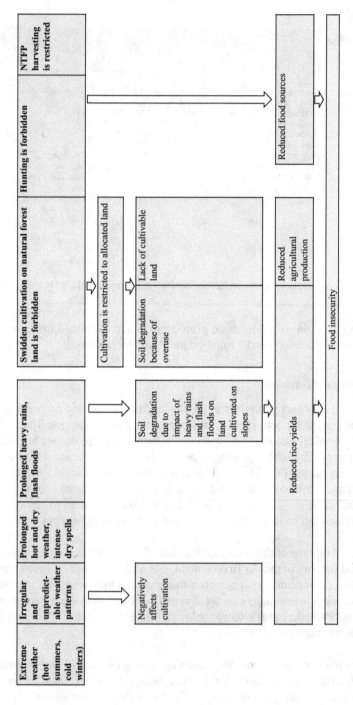

Figure 4.2 Impacts of climate change and regulatory changes on food security

The current rice variety of the local communities gives such a low yield that it can no longer meet people's food needs for the whole year.

(IDI with a 43-year-old man, an agricultural extension worker)

The results of the household survey conducted for this study, comparing the 1980s with 2012 (when the survey was carried out), also support the interview data in that only 38% of respondents surveyed reported that their households could produce enough rice to last a 9–12 month period in 2012, whereas 68% said that they had been able to do so back in the 1980s. Moreover, 43% of respondents said they could now produce only enough rice to meet three to six months of the household's rice demand, in contrast to a much smaller 9% who were reported to be in this situation in the 1980s. Finally, 10% said that in recent years they could produce enough rice to meet less than three months' worth of a household's rice demand, which is double the number (5%) who said this regarding the 1980s, as presented in Figure 4.3.

Key informants, such as an agricultural extension worker, also confirmed the reduction in rice yields, a pattern that has continued to the present:

We have been experimenting with new rice varieties for local communities in Nam Giang District. However, we have not been successful so far because all of the varieties [we tried] were not resistant to pests and diseases and so were not as good as the current one that local communities are using. We are still looking for a better variety that will be resistant to pests and diseases and also gives higher yields. The current rice variety of local communities now provides a much lower yield than in the past, and it can no longer meet people's food demand for the whole year.

(KII with an agricultural extension officer working for the Department of Agriculture of Nam Giang District)

In addition to reduced rice yields and agricultural production overall, restrictions on trapping and the prohibition of hunting, both results of the forest protection policy, added to the community's food insecurity as the Co Tu villagers lost one of their main sources of food. Although meat from hunted animals did not play a big part in the Co Tu's daily diet, it was important for the community's New Year, weddings, funerals and other occasions. Furthermore, the harvest of non-timber forest products (NTFPs) has also been reduced because the Co Tu people are no longer allowed to freely collect NTFPs in natural forests, including the bamboo shoots, honey, wild vegetables and fruits that were traditionally a food source to Co Tu villagers. The collection of rattan and malva nuts, which could be sold to outsiders, has also declined due to unsustainable over-exploitation.

After enforcement of the regulations began, there was some initial discussion of how hunting and logging continued in ways that clearly violated the new regulations (the products of these activities were often sold to outsiders). However, it appears that stricter enforcement has had an effect and that these activities are

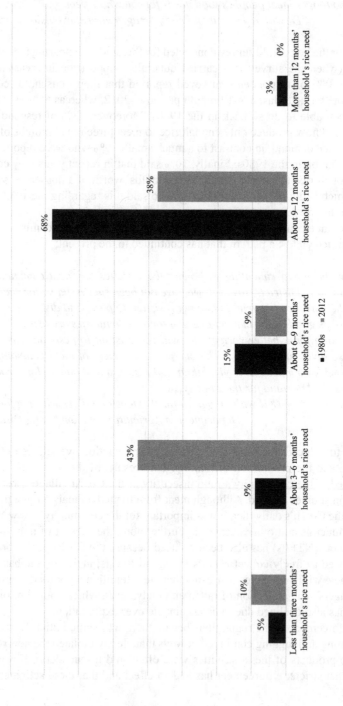

Figure 4.3 Percentage of respondents meeting rice needs in the 1980s vs. 2012

Source: Household survey, 2012, n=300

now highly restricted, particularly compared to the years before the regulations were implemented.

In sum, given the difficulties associated with rice yields, agricultural production overall and livestock raising in a changing climate, as well as the loss of food sources from hunting, trapping and NTFP collection and the lack of reliable alternatives at the present time, food security has become the biggest challenge and concern to local Co Tu households. Something had to be done in response, and the local women were the ones who struggled to make changes to increase food security – with important consequences for gender roles, relations and hierarchies.

Impacts of decreased food security on gender relations

As discussed previously, Co Tu villagers traditionally follow a strong patriarchal pattern with men as the head of the household, holding political power and making decisions that relate to the household and every member of the community. Meanwhile, women are responsible for looking after their families and being the key laborers who provide for the family's agriculture-based livelihoods. For this reason, Co Tu women were the most immediately affected by increases in food insecurity because, traditionally, they have been socially designated as the ones responsible for providing food for their families.

It is important to remember that rice cultivation continues to be the key survival strategy for local families. In spite of this, 72% of female respondents who were interviewed affirmed that their husbands would not worry too much if they ran out of rice because of the thinking commonly held, that "God made elephants, God made grass" (in other words, the belief was that because God let people be born, God would provide food for everyone). Women, on the other hand, would have to struggle on their own to ensure the family's survival through periods of hunger.

As a result of their role as key laborers in agricultural production, Co Tu women have access to the household's land and resources, but they do not own the family's assets. According to the customary law of the Co Tu, women generally do not inherit their *natal family's* assets, which include the house, land and other valuable possessions. This practice continues today; as an example, one of the male respondents noted that he inherited all of the family's land and assets, even though he has five sisters.

Moreover, as a reflection of the patriarchal structure, Co Tu women generally do not have control or power over their *husband's family's* assets either, according to both customary law and state law. State marriage law stipulates that in the case of a divorce, assets that are purchased *after* a couple is married will be equally divided. However, in a forest-based agricultural community such as this, valuable assets, such as land and houses, that are established before marriage are generally acquired through inheritance. Therefore, in rural agricultural areas where the land was completely allocated to households many years ago and women join their husband's household, a woman could end up getting nothing if she chooses to leave her husband and in-laws, given that the assets were not purchased after the couple was married. Because she is not likely to have an inheritance from her natal family (and she would generally not be encouraged to return to her family's home), and because she would not be likely to have access to a home or land through her husband's family after a separation or divorce, a woman would have difficulty even being granted care of her own children because she could not support them. (We should note that the subject of divorce is rarely discussed in the community, and it is not common for women to leave their in-laws' household.)

In addition, given the remote and isolated history of this community, state law would still not have as strong an impact as the customary law that has been guiding the Co Tu's cultural and social life for generations. Reflecting the patriarchal system, customary law states that a woman is subordinate to her husband and her in-laws. According to this cultural system, a woman, after marriage, becomes her in-laws' "person," as in the common saying "live as the in-laws' person, die as the in-laws' ghost" (i.e., even in death, she belongs to her husband's family). Customary law thus reinforces the perception that a woman would not have any place to go if there are problems in the relationship with her husband and in-laws.

Under such an unequal gender system, women's and men's roles and statuses in the Co Tu community are accordingly unequal in terms of the division of labor and in decision-making processes as well as power relations within the family. As discussed in the previous chapter, due to the major changes that came with the new land use and forest protection laws since 1991, women's and men's lives and livelihoods have been visibly affected at the grassroots level by new regulations regarding the use and distribution of resources as decided according to state and customary law (Adger, 1999; Beckman, 2011; Chaudhry & Ruysschaert, 2007). As a consequence, women in particular have had no choice but to find new strategies to ensure their family's survival, given their economically and socially vulnerable position; they were charged with meeting their family's daily food and other basic needs and had little choice but to find a way to do so.

The following section will discuss the ways in which women in the community responded to these negative impacts of climate, regulatory and socioeconomic changes. It will also detail how their responses were shaped by women's and men's roles and statuses in their families and in the community.

Co Tu women's responses to increased food insecurity

Given that in Co Tu households women have socially and traditionally been assigned as the primary agricultural laborers and food providers, they were among the first people who noticed that irregular and unexpected changes of climate patterns were having a negative effect on rice farming. The Co Tu women experienced the most difficulties created by crop losses due to a dry spell coming earlier than expected or to prolonged periods of heavy rain or flash floods that came before the harvest was finished. They were also the ones whose agricultural livelihood activities were restricted to certain pieces of land whose soil quality had degraded due to regulatory changes.

For these reasons, Co Tu women were also the first and most active agents in responding to food insecurity, as presented in the following figure (Figure 4.4).

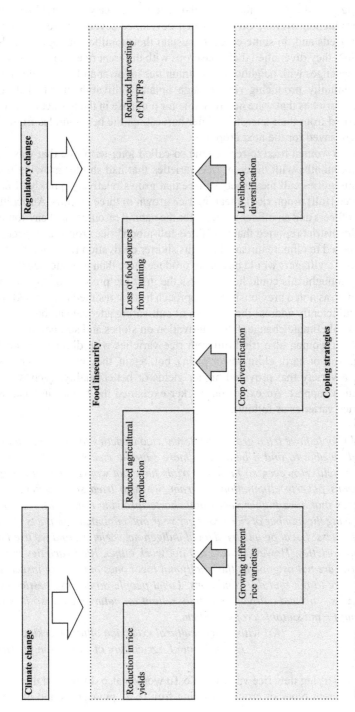

Figure 4.4 Women's coping strategies to respond to food insecurity

In coping with the lack of food caused by decreased rice yields and lower agricultural production overall, Co Tu women developed several strategies to meet their food needs and, in some cases, to sustain their families through periods of hunger. First, they diversified their rice crops with different rice varieties obtained through exchanges with neighboring communities in lower and higher elevations. Every community producing rice through upland cultivation has had its own upland rice varieties that were not available for purchase in the market but instead were inherited from their ancestors. After harvesting, the best samples from each variety were saved for the next crop.

The Co Tu women tried to replace the so-called long-term rice that was cultivated in six months with different rice varieties that had shorter growing times. Local communities call the variety of rice that grows relatively quickly "Lúa ba trăng" (three [full] moon rice), meaning rice grown in three months. Agricultural extension officers and an officer working for the natural resources and environment office in the district reported that the "three-full-moon" rice crop was expected to be less exposed to climate threats due to its shorter cultivation time, and the hope was that Co Tu villagers would be able to produce more than just one rice crop per year. They thought this could help address the growing problem of food insecurity; however, as noted previously, this approach has not resulted in the yields necessary to sufficiently address the problem of cultivating adequate amounts of rice in a context of climate change, fixed cultivation on slopes and soil degradation.

The Co Tu women also tried different rice varieties with different characteristics (e.g., soft or hard, glutinous or plain), but, again, they were not successful in finding a variety that provided higher yields or better quality, even with the government's support. An extension worker explained the failure in trials with different rice varieties as follows:

> *Local people have tried growing different rice varieties in the hope that they would be able to find a better and more suitable rice variety that gives a higher yield. However, so far, these trials have not worked. Under the government's poverty elimination program, we also tried some different rice varieties that have higher yields and shorter growing times with the aim of increasing the number of rice crops per year and minimizing impacts of natural disasters. Local people tried local indigenous varieties, and we tried new hybrid varieties. However, these did not work either. Rice varieties tried in this way are not as good as the traditional local ones in terms of insects and resistance to hot and dry conditions. Local people do not use pesticides or fertilizers – upland cultivation is fully reliant on nature's care, so it is very difficult to find suitable crops for them.*
>
> *(KII with an agricultural extension officer working for the
> Department of Agriculture of Nam Giang District)*

In addition to trying new rice varieties, Co Tu women also diversified their farming with different types of crops, including fruit (e.g., pineapples and bananas), beans, peanuts and lemongrass as well as their more traditional crops, including cassava and corn. Among new crops introduced in the area, the women's bean crops (green beans and black beans) have played an important role. According to the commune's agricultural extension officer, this strategy was successful in that the new bean crops were able to reduce household poverty in many cases, at least as long as bean prices remained high. During the years of high prices, the bean crops brought in the biggest and steadiest source of cash income for about 70% of Co Tu households, according to the head of Ca Dy Commune. Generally, with each bean crop, one family could earn about 1.5 million–3 million VND (75–150 USD), depending on their land availability and the number of family laborers who could help with bean cultivation:

> *I have not been able to produce as much rice as I did in the past. Rice yields have been significantly reduced, so I have been growing other crops to sell for cash, such as beans and ginger. In recent [good] years, we could sell beans easily for a good price. We could earn about 2 million VND for each bean crop. This money was saved to buy rice or cover other expenses.*
> *(IDI with a 43-year-old woman)*

However, what has happened in the last few years is that there has been less market demand for beans, and prices have fluctuated every year. This put the local Co Tu producers in a vulnerable position again, as explained by one of the women farmers:

> *The prices fluctuate all the time. This year [2019], they are dropping again. And it is very unpredictable – if we have a good year, we would make 10 million VND [432 USD] per season. In a bad year, when bats eat up everything, we're lucky if we can make 2 million [86.4 USD]; otherwise, it's difficult to even make 700,000–800,000 VND [about 30–34 USD]."*
> *(FGD with a group of women farmers)*

In the new cash economy, the villagers have also remained dependent on traders who come to the commune, and because of their mobility, the traders have much more experience and information compared to local villagers. This may help explain the fluctuations in the price of beans and other products and the problems Co Tu women face when prices end up lower than expected, which is something they could not anticipate without more information. Apart from limited capacity in accessing market information and negotiating prices, the Co Tu community has also lacked the technical skills needed to deal with pests, crop diseases and soil degradation; they also have lacked access to the cash that can help them purchase inputs to control the crop's quality. Agricultural cultivation is now mostly "left to nature's care," with high hopes and expectations for "God's mercy." This is why the ceremony held at the beginning of every rice season, in which they pray for a good crop, is a centrally important event for the local Co Tu community.

Co Tu men's responses to increased livelihood insecurity

Apart from *food* insecurity, the Co Tu women and men have also been facing an increase in *livelihood* insecurity caused by climate and regulatory change. As discussed in the previous chapter, changes in land use and forest protection policies have had extensive negative impacts on the Co Tu women's and men's livelihoods. The Co Tu men lost their most important livelihood activity – hunting – along with strong restrictions on trapping, NTFP harvesting and other forest-based livelihood activities. Unlike clearing the land to prepare for a new crop, which is now done by women since the big trees are gone, hunting has always been socially designated as strictly a man's task; it was thought that women can cut trees, but they can never go hunting since hunting is a reflection of one's masculinity in this community (Luu, 2007).

The loss of hunting has affected both the Co Tu men and women in negative ways. First, the Co Tu men have experienced the loss of their passion and their key livelihood activity. Second, the women in the family have had to cope with reduced food sources (and sources of income when traded) caused by the loss of men's livelihoods. In addition, interviews carried out in 2011–2012 and 2014 suggested that the fewer livelihood activities that men are responsible for, the more time they tend to spend sleeping, drinking or doing nothing at all.

The loss of their traditional activities clearly had the effect, for many Co Tu men, of lessening their interest in or involvement with providing for their family's livelihood. Interestingly, other men outside the community had similar observations about the Co Tu men's involvement in their family's income-generating activities. As an example, an agricultural extension worker who worked in the area (a Kinh male respondent) noted that the local Co Tu men had become less engaged in earning income for the family, particularly since they were no longer able to participate in hunting and other forest-based activities. According to the extension worker and other observers, Co Tu women have had to work harder than ever in order to provide food for their families.

Many of the Co Tu men continued to describe their main livelihood as farming and said that they work in the fields with their wives. "I practice farming. Every day, I work in the fields with my wife. In the past, I did hunting, but it is no longer permitted, so now I just grow rice. Whatever my wife does, I join her," said a 46-year-old male respondent in an interview conducted in one of the early visits to the community. However, this seems to conflict with the description of most of the Co Tu men's activities as given by female respondents and Kinh male respondents. The research team found, in visits to the community in 2011–2012 and 2014, that the presence of men at home in the middle of the day and during working hours was common. One female respondent who worked long hours in the fields to feed her entire family, including her in-laws, while her husband and her brother in-law stayed home drinking or sleeping, supported this observation that men were no longer as engaged in providing for the family:

> My husband's family was almost broke after our wedding. That's why I think
> I have to work to feed my entire in-law's family because I am responsible to

pay for the debt. Even my mother would not support me if I had a conflict with my in-laws, although sometimes I feel very unhappy because nobody helps me with farming while my husband and his brother do nothing at all, all day long.

(IDI with a 47-year-old woman)

Thus, we find that in the years following the sudden changes that happened in this region, it was evident that men tended to accept their new circumstances passively (i.e., without clear active responses to their new circumstances). However, because women did not have the option of a passive response, they took up the challenges as best they could.

Women's responses to livelihood insecurity

Women were not as directly threatened with livelihood activity loss as men were, but they had to adapt to new practices. Moreover, during the years in which Ca Dy Commune lost the government's subsidies for poor communes (2008–2012), Co Tu women were under a great deal of pressure not only to feed their families, but also to pay for children's tuition fees and meet other unexpected cash expenses. As noted earlier, in response to the shortage of money to cover these expenses, women diversified their livelihoods by engaging in new cash-oriented livelihoods. Some women engaged in small-scale trading, while others sold agricultural products in the market or to middlemen (including beans, corn, cassava or the remaining NTFPs they could access, such as rattan, malva nuts and Bon Bon fruit). The following discussions regarding ways to increase cash incomes are from interviews conducted during the first study visit:

To earn more money to buy rice and pay for the children's tuition fees, I grow beans and lemongrass for sale. I also have tried raising chickens, but this has not been very successful so far.

(IDI with a 46-year-old woman)

I think bean crops and malva nuts are the most profitable activities now. These products can be easily sold with good prices. They [traders] often came to our village to buy malva nuts. However, this has become much more infrequent nowadays.

(IDI with a 38-year-old woman)

We are often short of cash. My daughter had to quit school because I could not afford to pay her tuition fees. I do not know what to do to earn more money. There are no job opportunities for us here. They would not want to hire us for seasonal paid jobs, such as construction or gold mining – they would prefer men, as men are stronger.

(IDI with a 48-year-old woman)

Although women were often willing to take any kind of job as long as it increased their family's income, it has been very hard for women to find jobs that allow them to take care of both productive and reproductive work. This is partly because upland farming is both very time and energy consuming. However, it is also due to the fact that, for a long period of time, the scarce jobs that existed were generally more accessible to men, particularly to members of the Kinh majority rather than to members of the ethnic minority community.

This lack of alternatives for local men may also help explain the patterns of drinking, sleeping and idleness among Co Tu working-age men who could not find new ways to engage in livelihoods seen as appropriate for men after their earlier activities were prohibited. One question that emerged at this time was: How long could this pattern of idleness go on? How could men retain their status in the family and community when, in fact, their contributions to their families and the community as a whole were clearly decreasing, particularly relative to women's contributions – which were diversifying and increasing? A few tentative answers to this question have started to emerge, but long-term answers remain elusive, as will be explained next.

Changes in women's and men's livelihoods and the division of labor

Along with the Co Tu's traditional gender roles, the community's highly patriarchal gender division of labor reflects the gender hierarchy. Table 4.1 represents the traditional and more recent gender divisions of labor, based on the first field visits to the commune. As noted in this table, traditionally, both women and men were involved in upland rice farming; however, because regulatory changes have reduced or eliminated many of the tasks done by men, women are now the primary workers in upland rice farming. As a local Women's Union officer observed:

> There are no more big trees to cut now. Women can take care of cutting trees to start a new crop. For this reason, men do not participate in farming as much as they did before.
>
> (KII with an officer working for the Women's Union at the commune level)

Apart from the decrease in men's involvement in farming, men's participation in other livelihood activities was also found to be sharply reduced, especially in the years following the regulatory changes. The following quote, from an IDI with a 49-year-old woman, implies that some men were not very active, except when occasionally trapping animals such as small pigs and small porcupines (which, although not allowed, has not always been strictly monitored):

> They don't do anything. The old man [father-in-law] does some work with my mother-in-law to make baskets and carriers worn on the back. Every day, my husband and his brother sleep late and then go to pass the time with other

Table 4.1 Changes in the gender division of labor in response to food and livelihood insecurity

Livelihood activities	Gender division of labor in the 1980s		Gender division of labor after regulatory changes (based on the 2012 study visit)	
	Men	Women	Men	Women
Rice farming	Cut trees for land preparation and harvesting	Took responsibility for seeding, weeding, maintenance and harvesting	Partially and occasionally involved	Fully in charge of seeding, weeding, maintenance and harvesting
Hunting	Fully in charge	Not involved	Occasionally go for trapping	Not involved
Livestock raising	Rarely involved	Fully in charge	Rarely involved	Fully in charge
NTFP collecting	Partially and occasionally involved	Mainly in charge	Mainly involved in malva nut collecting	Mainly in charge of NTFP collecting
Selling agricultural products and NTFPs	Rarely involved	Occasionally involved, as this is a new livelihood activity	Not involved	Fully in charge
Bean and other short-term crops	Not involved	Occasionally involved, as this is a new livelihood activity	Occasionally involved	Fully in charge

Source: FGDs with female and male groups and data collected from the Commune People's Committee, Ca Dy Commune, 2012

men in the village. Sometimes they go hunting if they feel like it. Every now and then, they catch a porcupine or a small pig. Then they will invite other men for drinking.

What, then, were the men mostly doing during the years following the rapid changes described earlier? This question was asked of women during the early study visits (2011–2012 and 2014). Although not many of the Co Tu women respondents said explicitly that men in the community "do nothing," their answers about their husband's work or roles in household livelihood activities were vague, with no specific tasks identified as being performed by men. Apart from some women respondents who said that their husbands did not engage much in household livelihood activities, other respondents would say that their husbands were involved in farming. However, during interviews, when asked what specifically their husbands were doing, or what their husbands were responsible for with regard to farming, many of the women respondents had no answer. When asked

"What does your husband do when you run out of rice?" a common response from wives was, "Women have to take care of daily food." There was a saying that was repeated quite often in the interviews about husbands' and wives' responsibilities, to the effect that husbands take care of "big" things while wives take care of "small" things; however, in reality, a gender division of labor along these lines was no longer possible because "big" things, such as cutting trees and hunting wild animals in the forests, were no longer permitted.

Thus, in the years following the enforcement of regulatory changes, there was a large gap between Co Tu women's and men's efforts in coping with the impacts of the changing climate and regulatory policies on food and livelihood insecurity. These are the changes, in the end, that have challenged unequal gender relations and, in turn, the entire patriarchal system. Women's frustration was often voiced during field visits to the community during these years, and it was not clear how the impasse of men not taking more active roles in providing for their families could be resolved.

Relationships between husbands and wives, and changing roles in the household

As discussed previously, men's roles and contributions to their families were altered dramatically following the imposed changes of the past two or more decades, while women consistently took on key roles in providing for the family's income and wellbeing. Along with these increased burdens, women's roles and importance with respect to their families and communities – the latter particularly when dealing with traders and other outsiders – also increased in new and significant ways, which could be seen as an indicator of women's empowerment. Before all these changes, the Co Tu villagers had followed an extremely patriarchal pattern. Women could not make decisions, even ones related to their own personal matters. For example, their participation in government-sponsored trainings, community meetings or other public activities was limited, as they were not encouraged to participate in such non-household-based activities because it was said that their husbands did not want them to associate with other men. Additionally, and perhaps more importantly, community and public activities were not considered suitable for women, as they had traditionally been activities assigned to men.

This changed when women's roles in providing for the household's livelihood needs increased and men's roles decreased, as described in the following quotes from the first two study visits:

> *In recent years, my husband has been much easier to me [more flexible] than he used to be in the past. He lets me participate in village meetings, which he did not support a few years ago because he said it was not women's business. I think he has changed because I have been working very hard to take care of our family and his parents.*
>
> *(IDI with a 45-year-old woman)*

Co Tu men are very jealous. They would not want their wives to interact with other men. Years ago, Co Tu women did not appear much in the public sphere like they do now. Things have changed a lot, I think, over the past ten years due to a series of trainings that were conducted under the government's poverty elimination program. Women are the key agricultural laborers, so men have to let their wives participate in trainings for their own sake.

(KII with an officer working for the Women's Union at the commune level)

Since women started becoming more engaged in public activities, such as trainings provided by the government regarding farming techniques, agricultural production-related decisions (e.g., crop selection) and household financial management and savings, men got used to women's increasing interaction with others beyond the household and with the outside world in a more general sense. Increased social interaction and participation in public activities have contributed to reducing women's isolation in the community. Although it has not yet created significant changes in terms of gender relations that are explicitly recognized as more equal, these activities have broken the traditional image of women as being someone's wife or someone's daughter to one in which women are seen as independent individuals with their own personal activities and their own social networks.

As part of this process, women also started seeing themselves as individuals with their own rights and opinions, as the following quote suggests:

Things have changed now. We make decisions altogether. Since I work hard in the field, my husband also helps with the preparation of dinner if I come home late.

(IDI with a 43-year-old woman)

The women also started to understand how their relationship with their husbands was unequal:

I don't like the fact that my husband is acting too lazy and does not help me at all to provide food for our family. However, I cannot do anything about it because it has been like this since our ancestors' time. Even my mother would not support me if I fight with my husband. All the other women in my village are in the same situation.

(IDI with a 47-year-old woman)

Questioning the inevitability of the gender hierarchy in this way can be seen as one of the very first steps Co Tu women are taking toward moving away from unequal gender relations. In addition, opportunities such as government-provided trainings and meetings for livelihood improvement have also helped the Co Tu women develop their sense of self-confidence, encouraging their self-reliance as well as expanding their social networks. In the long run, this can help eradicate the local perception of women as being inferior to men.

Although often shy, female respondents who were interviewed were not very difficult to reach because many of them seemed to be interested in joining interviews, especially FGDs (group interviews). This research benefited a great deal from their willingness to participate in interviews and other research activities, as revealed in the following answer by one respondent after the lead researcher apologized for the long interview with her:

> *It's all right. This is fun. We don't often have activities like this. This is also an opportunity to get together with other women to talk and share. I am glad that I can help. Just ask whatever you need to ask.*
>
> *(FGD, 53-year-old woman)*

This quote may also reveal the fact that farming is often lonely work for women. Given that farming activities in most cases have been almost entirely handed over to women, the women sometimes talk of walking for long hours and working alone in the fields. Much of the work that used to be shared between husbands and wives has been significantly reduced, and after land was allocated to individual households and collective activities overall in the community were reduced, the focus on individual production appears to have had the unfortunate consequence of discouraging group coordination and cooperative practices. This can leave women and other members of the household more isolated than before, given the traditional emphasis on household production and village-centered activities, rather than production carried out by each person separately, along with the new emphasis on personal consumption and other individual-centered activities.

Still, the changes that have come to the community are complicated and certainly not all negative in nature. As discussed previously, increasingly over the past two or more decades women have entered the cash economy and often engage in small-scale trading to cope with food and livelihood insecurity. Cash income from selling cash crops as well as some NTFPs has brought an important change to the lives of families that did not have a regular cash income 20 years earlier. With money in their hands, women also feel more in control of taking care of and providing food for their families. One woman said the following:

> *Since being able to sell things like malva nuts and rattan collected from the forest, or beans, peanuts and cassava, I feel much better as I have money. It is not a lot, but it makes me feel more secure when I have some money. That's why I started saving money little by little. This will help my children finish school.*
>
> *(IDI with a 39-year-old woman)*

With an increased cash income, Co Tu women are now able to meet many of their family's needs and are less worried about providing food for their children and paying for their studies (this was particularly important during the time the commune lost its "especially poor" designation):

Since cultivating the bean crop, we now have cash more regularly. In recent years, we have started selling things in the market. We sell malva nut and Bon Bon that we collect from the forests during summer months. We sell beans, peanuts, and other agricultural products that we have left over after our own consumption. I often save this money for rice-deficient days and to cover children's tuition fees. Every now and then I can buy some necessary things for the family. I also bought some chickens to raise. Last year, I bought a pig. Hopefully by the end of this year, after selling the pig, I can earn some money for Tet [Vietnamese New Year].
(IDI with a 44-year-old woman)

Co Tu women's roles and relative status have not just increased inside their households – their community has also started taking them more seriously. During FGDs, participants noted women whose status had changed a great deal; for example, one male participant stated the following about a woman who had just left the discussion:

She is very good at doing business. She is rich now. Her husband obeys her very much.
(IDI with a 27-year-old male respondent)

In the Co Tu traditional culture, "obeying" a wife would be shameful to a husband. This quote showed not only the admiration but also the envy of some members of the community toward women who adapted well to the situation and made the best use of the new opportunities. This quote also suggested a fear shared by male respondents regarding the fact the women started taking control while men started losing their status inside their families, and it revealed potential conflicts between women and men in their households.

In contrast to the Co Tu women's increasing roles and status, men's roles in their families decreased along with their workload and responsibilities. Two to three decades ago, when men were more involved in the household's farming as well as hunting and other livelihood efforts, they took charge of household activities and participated regularly in meetings and events. However, after the new regulations started being enforced, men showed a lack of interest in joining meetings or trainings due to their reduced involvement in family livelihood activities. Having more idle time caused an increase in men's drinking, and alcohol was not expensive in the study area, especially considering that the Co Tu villagers could also make their own alcohol. Even during the time of the research team's first two visits to the commune, it was common to see men drinking during the day while women were found working in the fields, as described here:

Men drink a lot. Right now, at this moment, you can see a group of men drinking in the opposite house. Before, they drank less. They could go hunting and

> *trapping for days or even a few weeks during the hunting season. In recent years, ever since hunting was forbidden, they spend a lot of time just staying home and getting together with each other. Young men play games. Middle-aged men drink.*
>
> *(IDI with a 43-year-old woman)*

Men's drinking habits and decreased involvement in household-related activities were also revealed in the following tragic story:

> *I think nowadays men drink more than they used to in the past. They have more time because hunting is forbidden. They also engage less in farming because farming is easier than it was in the past when we cultivated on old forestland. And, also, people now have more money to buy alcohol because they sell their agricultural products. One of my neighbors lost two sons to drowning. Their mother worked late in the field on the other side of this river. I was cooking dinner for them as I always did when their mother came home late. Their father was drinking. The two boys went to the riverside to call for their mother and drowned. It was so tragic that I could never forget. . . . The father died a few years later because of a liver problem.*
>
> *(KII with an officer working for the Women's Union at the commune level)*

It should be noted that there were no reports of an increase in domestic violence in the community (there may have been cases of domestic violence, but there were no reports of increased domestic violence since the changes discussed earlier started having serious impacts on the community). However, the way women complained during interviews about men drinking and being lazy clearly indicated the increase in tensions in the family since these changes took hold. Also, according to Bayrak et al. (2013), alcohol abuse was identified as one of the main problems, in addition to poverty, climate-related problems and land-related conflicts, in the Co Tu communities in the Central Highland Region (adjacent to Quang Nam Province and sharing much in common with the present study site). These problems were said to result in conflicts among villagers, conflicts between villagers and higher authorities (including the patriarchs or elderly) and also domestic violence, according to this 2013 study. Elder villagers say that in the "old days," the Co Tu men drank less alcohol than in recent years, and definitely less than middle-aged and elderly men have been consuming ever since these major changes started to occur.

The men's drinking and other behavioral changes undoubtedly reflect their changing gender-related roles and responsibilities, as the men most likely came to feel that their position and their own worth was being challenged. On the other hand, along with increasingly unequal gender-differentiated workloads and greater burdens placed on women, the women's involvement with the cash economy and central role in new livelihood activities resulted over time in a greater valuing of women's contributions to their families and to the community. Such changes helped to narrow the gap between women's and men's statuses in the family, and an increase in the women's role in earning an income also enhanced their voice and their participation in decision-making processes in the family and community. In this way, gender roles and hierarchies have started to become more equal, but at the same time we can see that this has come at a significant cost, to both women and men in the community – to men in terms of their declining status, contributions and perhaps self-respect, and to women in terms of heavier burdens and increased tensions within the household.

The younger generation and gender-differentiated expectations

Changes in the younger generation have also been taking place that need to be better understood, as gender roles and hierarchies in the family and community are being questioned in this way. In Co Tu families, for example, the education of boys used to be given higher priority than the education of girls. A girl child, for the most part, was expected to leave her studies before high school and get married. As one respondent explained during the first study visit:

> *My daughter dropped out of school after she graduated from the ninth grade. She will be getting married soon. Girls do not need to study to a higher level. They need to get married – otherwise, they might be left on the shelf [they will not be able to get married]. My son is studying in high school. We hope he can get a job and will not have to practice farming like we do.*
> *(IDI with a 37-year-old female respondent)*

Particularly during the time when the community lost government subsidies, many students dropped out of school, with a higher percentage of girls dropping out than boys. A boy child used to be encouraged to finish high school and go on to some form of vocational education with the hope of getting a paid job that was considered better than farming. Viewing farming as hard and of low status (and, again, it was seen largely as "women's work" by this point), young men and their parents dreamed of the young men getting away from working in the fields. For this reason, a boy child was not expected to contribute to his family's agricultural livelihoods and, consequently, boys did not learn much about farming unless the family was too poor and they had no other choice:

> *I do not like to do farming as the income is very small, and it's not worth the hard work that we put into it. I am studying in town and will find a job there.*
> *(IDI with a 20-year-old male respondent)*

> *My son is really a good boy. His father died young, so he loves me very much. He does not mind doing anything to help me earn more income. He does not play games or hang around all day like many other young men in the village. In recent years, some families dream of having their sons get away from farming. However, there have been very few of them who can actually get a*

non-farm job because our children don't study as well as people do in the lowlands. Most young men, coming back home after studying at high school in town, are now unemployed and don't know how to do farming. I think we are farmers. It would be better if we were good at farming.

(IDI with a 59-year-old woman)

At the time of the first visit to the study area, many young men in the community were jobless because they were not good at or interested in farming, but at the same time they were not able to get jobs elsewhere because they could not compete with people from lowland areas. This was said to be due to the fact that those from lowland areas had better social and information networks, better educational backgrounds, more available capital and better communication skills (according to KIIs with government officials). Given these obstacles, many of the young Co Tu men appeared to be unmotivated – staying at home, not helping their families with farming and just waiting in the hope of getting a paid job in the lowland areas. Meanwhile, they also did not contribute much to their families' income, as learning about or doing anything associated with agricultural work was outside of their interest. They preferred to meet with friends to play video games while their mothers and sisters worked in the fields. In the meantime, their families still have had to support them both in terms of their study-related expenses as well as their personal expenses.

As indicated earlier, at this time families did not prioritize educating their girls as the young women were expected to drop out, return to farming and raise families. Those interviewed in 2011–2012 and 2014 did not suggest that young women spent time on personal hobbies or waited for jobs to open up for them in lowland areas. Instead, they were expected to follow their mothers' established patterns, which did not involve going far from home or taking up new types of occupations.

Gender relations in transition: more recent developments

In recent years, there have been some notable changes in both women's and men's efforts to cope with food and livelihood insecurity. This has included modifications in the behavior of middle-aged men away from the patterns of idleness and disengagement; however, many of those interviewed indicated new problems that still remain to be solved; these new developments will be discussed further in this and the following chapter.

Again, one important question that was raised after the early field visits to Ca Dy Commune in 2011–2012 and 2014 was: How long could this situation of men being idle and losing their identities as contributors and women being overburdened last? Certainly, there are large numbers of cases all over the world of men sitting idle – drinking and talking with each other – while women work very hard, and it appears that in many places men are able to sit on top of gender hierarchies without feeling they have to change in any way. However, in this case, the sudden elimination of what men saw as their primary roles in the family and community brought up the question of how long they *could* remain idle, given that they had been active and that their earlier contributions had given them a purpose and sense of self-esteem, at least until they could no longer carry out those designated roles.

Other factors that appeared to push toward change included the fact that during the early visits to the community, conflicts within the household were said to be increasing, in part due to the perception that women's status was improving because women were the ones engaging in the new cash economy through their agricultural and other cash-generating activities. The question was, then, how far could this change in gender roles and hierarchies go – would women become the leaders in their families and community over time, given their status as the clear source of livelihoods in a community threatened by both climate change and a rapidly changing regulatory and socioeconomic context? Or would men feel internal pressures as well as external pressures and incentives to change their behavior and find new ways to adapt to the entirely new conditions they faced?

Recent changes affecting middle-aged men and women

Interestingly, nearly ten years later, answers to these questions – which are still evolving – have started to emerge. It appears that there are indeed limits to the number of years men would want to stay idle, outside of the cash economy, while women were adapting as best they could to the changing circumstances. Specifically, by the time of the 2019 field visits, it was found that middle-aged men had begun to work increasingly as day laborers, such as by doing daily wage types of work in acacia plantation fields or in gold-mining operations. These are unstable and insecure jobs; they are temporary and seasonal. However, they do bring in a monetary income and bring the men closer to the cash economy that had been primarily the realm of women, as the women adapted to meet the new social and environmental context and the new requirements.

The men's movement into daily wage labor represents both a change in attitude and behavior on the part of middle-aged Co Tu men in the community. It was noted in interviews that older men in the community often did not mind working alongside women by engaging in agricultural work or in making baskets and helping with other useful tasks. This may have been due to the fact that the older men were used to contributing to the family's livelihood, had gone through the traumatic war years, had already established their status as well as secured their role as men and their sense of masculinity, and did not feel required to prove their masculinity to those around them. However, for middle-aged men, participating in what they considered to be work inappropriate for men (e.g., in agricultural work ["women's work"] under the present circumstances) was not something they were initially prepared to do.

Moreover, as one 42-year-old woman noted (quoted in Chapter 3), middle-aged Co Tu men were said to have considered working for other people as shameful ("he does not like to be seen as a hired laborer working for a boss"). This can be understood given their history of living in a close community where members felt they were like family and shared everything, rather than being seen as "workers" who have "employers" (interpreted as the relationship of an inferior to a superior).

In recent years, even the idea of working for other people has gradually become more acceptable – or even more than acceptable, it has become attractive, given the importance of earning cash in this new social and economic context.

In other words, it appears that the lure of the new economy has begun to change attitudes about what is now considered acceptable or preferable as "men's work." In interviews held in 2019, the majority of female respondents reported that their husbands have now periodically engaged in day labor, with seasonal and short-term incomes. From this, we can see that this cultural obstacle of not wanting to be seen as "hired labor" now appears to have been resolved somewhat for the majority of men. It is possible that even though the men originally felt uncomfortable with or resented the idea of working for others in this way, enough time had passed for the men to get used to the new livelihood options as they gradually came to the community; it is also possible that the men felt these short-term types of employment were still under their control, since they did not commit for long periods of time.

This change in attitudes may be partly explained by the perception of the need for cash, both as a contribution to the family and for the men's own use. The change may also have been tied to the boredom of having very little to do, or could in part be the result of new identity-related images (e.g., of "appropriate" male employments) and new opportunities coming in from outside the community. It is also possible that the men were aware of the tensions and complaints expressed by women and elders toward their idleness, and they may have been acutely aware that their status was eroding as the women of the community became more involved in the cash economy.

However, another important part of the explanation is no doubt tied to the availability of work, which either had not been available locally or was often given to Kinh (ethnic majority) men instead. Some of the new opportunities for men have been associated with the expansion of cash crops. Acacia is notable in this regard: It was introduced into the area more than 20 years ago, and many of the slightly better-off households that have a good deal of cultivated land (more than one hectare) have established acacia plantations and now hire day laborers from the ethnic minority community (i.e., Co Tu men) to do the work.

Acacia, in recent years, has also provided higher profits than other cash crops, such as cinnamon or rubber, since it can be harvested in five to six years, and a single household growing acacia can earn from 10 million to 20 million VND (about 432–864 USD) on average for each crop, which is considered a very large amount of money to the Co Tu in this area. In the case of day laborers, one person can earn about 150,000–200,000 VND (about 6.5–8.5 USD) per day. The problem here, as with other cash crops, is the instability of prices: Prices for acacia were good from 2010 to 2012, then declined in 2013–2014, and moved up again in 2017–2019 (i.e., by the time the most recent interviews for this study were conducted).

From the point of view of men working in the plantations, acacia has the additional benefit of not requiring constant care; it is not as labor-intensive as rice production and offers a great deal of flexibility. Apart from working on acacia plantations, the Co Tu men can also work for a few days as day laborers in other locations, then spend a few days at home and collect NTFPs (to the extent allowed) whenever they choose. Their livelihood preferences have thus changed over the past two decades from resisting work after being displaced from their earlier livelihood activities toward a willingness to work as day laborers as long as they have the flexibility to choose when, where and how much they work.

What is clear is that men in the community have now decided to enter the cash economy and, as a result, have become more active in earning income for their families as well as for their own personal consumption. One man explained why it is now important to earn a cash income (interestingly, he downplayed the key importance of women's arduous work in rice production, even though it continues to be a key survival strategy for poor Co Tu families):

> *It is not too difficult to find something to eat. Even if we ran out of rice, we could eat cassava or corn instead; every household has some cassava and corn in reserve [in case of rice crop failure]. Every now and then we could catch some fish or snails. What worries us most is money to pay for the kids to go to school or money to give for weddings and funerals.*
>
> *(IDI with a 37-year-old man)*

On one hand, this entry of men into the cash economy is a relief for women who were overburdened with being solely responsible for meeting the family's livelihood needs. On the other hand, the men's entrance into paid labor posed a new challenge, given that during the field visits in 2019 it was found that women's earnings had become much more uneven than in past years. Part of the explanation for this had to do with the continuing negative effects of the changing climate and soil degradation on agricultural production, but it is also tied to their dependence on market prices that can change quickly and in unpredictable ways. For example, beans currently do not earn the amount of money they did previously, and the women are therefore required to conduct an ongoing search for new cash crops.

Moreover, and very significantly, the women's *non-cash* work in rice production for the family's own consumption – critically important though this is for these low-income families – is now not given the same recognition as work that results in a cash income and is often dismissed as relatively unimportant ("God made elephants, God made grass"). In other words, as the community becomes more fully involved in the cash economy, the trend is that, increasingly, *only cash-generating activities* are valued as significant and worthwhile. Non-cash-oriented activities, such as rice cultivation for the family's own use, are not seen as desirable by men – and, perhaps in part as a result of this, increasingly not by women either, no matter how crucial these activities are in reality. (This may also be due to the fact that cash-earning activities buy things that are new and unusual, in contrast to activities that result in such "ordinary" things as rice and other goods for the household's consumption.)

Some recent small-scale development projects have also been cited in interviews as hurting rice production by reducing the rice yields the women had been able to achieve earlier, even under conditions of climate change and soil degradation. Specifically, in 2019, women reported that their agricultural cultivation has suffered due to new and unanticipated problems associated with livestock raising, in addition to severe weather and other challenges. They stated that over the past five or six years, cows, buffalo and pigs were provided to poor households by the government and private-sector enterprises (the latter as charitable contributions) toward the households' economic improvement. They reported that the number of cows, buffalo and pigs has gradually increased, and because the livestock are left to graze freely, there has been a strong negative impact on rice cultivation wherever livestock are found. As a consequence of these disturbances – climate threats, increasingly unproductive soil, lower and unstable market prices and livestock-related problems – many no longer consider rice cultivation, either for cash or non-cash purposes, to be a viable livelihood option, at least for those who are in a position to move away from rice production.

This idea was reinforced when rice and other short-term crops were often and repeatedly referred to by many respondents in 2019 as "not worth investing in." They reported that an increasing number of households (those somewhat better off) now leave their land idle or rent the land out to others if a family can find better livelihood options. If this move away from rice production continues in the future, it will represent a major change to the local Co Tu way of life and certainly a major change in livelihood activities for the women of the community. However, we should note that the great majority of households in this area still do not have other options; *they have no choice but to continue on in this way*, no matter what problems – tied to climate change or otherwise – emerge to challenge their agricultural livelihoods based on rice production along with other crops, at least for the foreseeable future.

Livelihood activities, gender roles, and gender hierarchies are thus in transition, and the outlook is not very clear. We can say that with men no longer being as idle as they were before, the workload and economic contributions of women and men have become somewhat more equal. Moreover, with the contribution of cash income from the men's day laborer jobs, the issue of women being overburdened and under constant financial pressures has been reduced, especially since the commune has been put back into the poor commune category so that government subsidies are again being provided.

However, even though both male and female respondents now talk about gender equality and have more awareness about the concept than was true earlier, a remaining – and in some ways more difficult – gender issue has emerged. Men appear now to increasingly desire to be seen as the primary cash-income providers while women remain tied not only to cash-earning activities but also to non-cash livelihood activities, such as agricultural cultivation for the family's consumption, and of course women continue to be responsible for the great majority of family and other household-related duties as well.

The problem is that these non-cash activities carried out by women, no matter how important, are not valued as much as cash-generating activities. Moreover, the women's cash-based contributions – particularly when they have so many other responsibilities – are declining relative to those of men who have few other responsibilities and can now devote much more of their time to earning cash incomes, even on a seasonal or temporary basis. This creates a possibility for men to earn more cash, and in so doing, they are likely to further devalue women's contributions relative to their own, in spite of the much greater time contribution and effort – including both cash-oriented and non-cash contributions – the women are actually making.[1] It is understandable that men who have had a higher status in gender hierarchies are reluctant to give up their privileges and sense of importance and may even unintentionally and unconsciously try to reassert their dominant position in whatever ways they can (as seems "natural" to them, given their earlier experiences). Even when a number of forces are pushing toward greater gender equality and a rise in women's status, the road can be full of advances and reversals, as seen in this case.

How these complex developments will affect gender roles and hierarchies in the future is unclear at this time. This will depend in part on the next generation of young women and young men, and it is to them that we now turn.

Recent changes affecting young women and young men

During the early visits to the community in 2011–2012 and 2014, there appeared to be a particular problem with young men who could neither fit easily into the older agricultural economy nor into the new cash-based economy. The young men were attracted to the "modern" economy and clearly desired its consumer products (mobile phones, motorbikes and the like). They may also have been given a sense of the relatively privileged position of men in the mainstream of Viet Nam's economy and society and were attracted to that as well. Like their fathers, they did not want to take up what was now seen as "women's work" in the fields, and their families invested in their education with the hope that they would enter the mainstream economy and find steady employment, rather than returning as farmers.

However, as discussed in the previous chapter (Chapter 3), these young men – coming from a rural ethnic minority that often did not use the national language at home – generally found they did not have the education, social skills, capital, connections or other attributes that would get them the jobs they desired. For this reason they were seen as "lost" (neither in one world nor the other), and respondents worried that the young men were wasting time and money and getting nowhere in terms of their family's and community's expectations in the "new" economy, as well as in terms of their own desires.

In the case of young men, this situation, unfortunately, has not changed much over the years. Even during interviews in 2019, teenage and young adult males, specifically those who were unmarried, were generally described as being lazy, unmotivated and as contributing virtually nothing to the families' incomes. One of the elders shared with the lead researcher his frustrations regarding the younger generation, including his own son; according to him, these young men tend to be irresponsible and lazy, have no willpower and do not seem to care for others, based on what he sees of their current behavior.

In earlier interviews, the young men's problems in their studies and their lack of connections were said to be the most important obstacles in finding a job. However, it was also noted in recent interviews that even if they were offered a job, many of the young men have shown that they may not want or be able to keep it. Respondents say that their inconsistent work habits and lifestyle, resisting regular work schedules that do not allow them the flexibility to stay away from work when the young men do not want to work, often prevent them from getting or keeping a job.

Specifically, respondents were concerned that a majority of Co Tu young men appear not to want factory jobs, or office jobs for that matter, with fixed working hours. As a result, many were seen by others in the community as doing very little apart from spending time with friends who also did not want, could not get, or could not keep regular jobs. In fact, this impression was reinforced in a group discussion with young men in the community, as it became clear they were not interested in jobs such as factory work that had tight and inflexible work schedules. It was also evident in Pa Don Village, where almost all of the seven young men who had been hired for factory work in a city in Quang Nam Province ended up quitting their jobs due to the same concern (which is perhaps understandable, given their previous experiences of having a good deal of unscheduled time):

> It is very hard, working for a factory in the city. Eight hours a day, and we cannot take leave. We cannot make any mistakes. Anything can lead to a deduction in our payments, such as being a little late or making a small mistake.
>
> (IDI with a 23-year-old man)

Interviews indicate that the pattern of not wanting full-time inflexible work is common until the young men are married. Once married, the young men who do not have "regular" jobs are said to generally adopt the pattern of middle-aged Co Tu men doing day laborer jobs – again, seasonally and temporarily, working only as much as they want or need to work, but earning cash that can be used to meet household expenses and also directed to the men's own use.

Interestingly, the situation of young women in the community is described as being quite different from that of young men at the present time. In the beginning of this process of rapid change, interviews conducted during the early visits to the community indicated that girls and women of all ages were affected most by the negative impacts on food security. An increase in the number of girls dropping out of schools, because they were expected to get married and support their families, was reported at that time.

However, some positive changes in young women's status have been observed in recent years. One reason is that, beginning in 2013, once the commune was put back on the "especially poor commune" list, fewer girls were made to drop out of school due to the reinstatement of government educational subsidies, as described in the following quote:

> Yes, yes, I recall that several years ago girls frequently did drop out of school. It was around the time we were not supported financially by the government.

I think it was around 2011, 2012 or 2013. But now, it no longer happens much because the government again supports us in terms of education. Nowadays, kids drop out of school mostly because they don't like to study, or they do not pass exams [to go on to high school or higher levels of education]. Girls now study even better and pass the high school exam more easily than boys do because they study harder.

(KII with the Chairwoman of Ca Dy Commune Women's Union)

In addition, due in part to positive influences from the community's new connections to the outside world, young women are now encouraged to study up to a higher level. Interestingly, interviews suggest that there have been even more young women than young men studying for bachelor's degrees and vocational training because more young women could pass the exams while young men have had difficulty passing them. The young men are said to be more distracted by video games and other activities they engage in with their friends, now that internet coffee shops have opened up both within and outside the commune, while young women are more motivated to do well in their studies.

In this way, even though both young women and young men still struggle to find stable and well-paid jobs in town due to their ethnic minority status and relative lack of skills and connections as compared to lowland or the Kinh majority population, young Co Tu women appear to be more successful in both their studies and in finding employment. It is true that their preferred jobs seem to be those considered to be most "appropriate for women" (e.g., teaching young children), even though these jobs are very difficult to get and relatively few can be hired for this type of employment. The young women are also reported to be willing to take on the types of regular and ongoing employment, with inflexible hours and few vacations, that young men resist. However, after marriage, a large number of the young women, particularly those from poor households, are likely to return to work in agricultural production on their husband's family's allocated land and will have to face the problems their mothers' generation faced.

It should be noted that, as the cash economy grows in Ca Dy Commune, there are also likely to be problems as well as opportunities associated with socioeconomic differences that emerge in the future. On one hand, the new economic and social forces may act to expand the range of possible gender roles and affect prevailing gender hierarchies. Respondents noted, for example, that one young woman from a relatively well-off family was able to become a medical doctor, which is a significant achievement for someone from a rural minority community. Moreover, as mentioned earlier, new ideas of gender equality are coming into the area, both through the work of such organizations as the Women's Union and also through examples of some young women achieving what many young men so far have been less interested in pursuing.

However, for less well-off (i.e., the great majority of) households at least, agriculture will remain their primary means of earning livelihoods, and the challenges faced by both young women and young men are likely to grow with an unpredictable climate and other new difficulties that can arise in unexpected ways. In the long run, harmonious relationships within the commune may be strained as well, as the community becomes less interdependent, less secure and more individual-istic in attitudes and behavior, with much more of an income hierarchy and much less of the social solidarity needed to face new difficulties as they arise.

We will come back to this in the next and final chapter, as we attempt to look toward the future and suggest ways in which the older, middle-aged and younger generations might be engaged in order to meet these new climate-related, socio-economic, policy-related and other challenges. For example, although national and local institutions, such as agricultural extension and other government and private organizations, have traditionally been oriented toward working with men and not women, the apparent aspirations and drive of these young women suggest that there are opportunities to be explored with both young women and young men from the community in the future.

This is particularly important, given that the fields and forests will remain the primary source of food and livelihood security for the local community, at least for the foreseeable future, and dealing with ongoing climate threats as well as rapid social change will be a crucial requirement for the community as a whole. Assisting young women and men in becoming educated in a way that will help them understand and deal with the new challenges, including their rapidly chang-ing natural and social environment, will be important. This will be true not only for meeting basic needs but also for bringing about more gender-equal and mutu-ally supportive relationships within the family and within the community as a whole. Different suggestions regarding how this might be achieved will be dis-cussed in the following chapter (Chapter 5).

Summary

This chapter began with the argument that a changing climate, together with a changing regulatory framework and changing socioeconomic and cultural pres-sures, has had major impacts on the Co Tu community's food and livelihood secu-rity. Moreover, it was found that Co Tu women and men have experienced these impacts in very different ways.

The findings suggest that in the wake of these changes, women's and men's roles in the family moved them in opposite directions. Increases in women's con-tributions to and roles in the family and community happened at the same time as decreases in men's contributions and roles. On one hand, this has forced women to work harder to the point that they became seriously overburdened, whereas men initially became idle, not wanting to work in farming or for other employ-ers as hired laborers, once their traditional livelihood activities were closed off to them. On the other hand, these changes acted to improve women's status, which in turn brought the prevailing gender hierarchy into question.

In short, the gendered impacts of the combined changes in climate and soci-ety have begun challenging traditional gender roles and relations among Co Tu people in Ca Dy Commune, resulting in increased tensions in many contexts but also providing new openings, particularly for women and girls in the community. Moreover, based on recent developments observed in the 2019 interviews, there have been some improvements in recent years for men as well as for women, but there are also significant new problems related to gender hierarchies and inequalities that have yet to be resolved.

The following chapter will begin with a summary of findings and possible contributions this study of a remote forested, mountainous community in Viet Nam may be able to make to discussions of climate change and gender. In particular, this study uses an expanded framework that has included other related driving forces that have compounded the difficulties posed by increasingly unpredictable climate extremes in a community that is dependent on the natural environment for survival. Based on these findings, and by considering different views regarding where to go from here, a few suggestions will be made in order to explore ways in which vulnerable and previously isolated communities – including the Co Tu community of Ca Dy Commune – might be able to cope more effectively with the negative impacts of major climate, regulatory and socioeconomic changes that have come together in such a forceful way and in such a very short period of time.

Note

1 This possibility was already indicated in Arhem and Binh's report from 2006 (p. 74), which quoted a woman in Dak Kroong Commune (Dak Lak Province, Central Highlands) that illustrated well how men in this context could elevate "men's work" and devalue "women's work" – including women's crucial contribution through agricultural production for both home use and the market, together with household work and other responsibilities – by defining "contribution to the family" as tied strictly to *cash earned*, rather than the totality of paid and unpaid work. In this case, men periodically earned more cash than women did (sometimes from unauthorized activities) when they weren't idle: "Here men just work if they like, otherwise they sleep all day or walk around the village. They only help women when clearing and sowing fields. Only a few men, who understand our hardship, help us carry rice during the harvest. Most just say 'that is your work' and then go and drink. If the women complain, they will say: 'how much do you earn from cassava and maize? On average you just earn about 2000 VND/day. *You live on me* . . . ' [our italics]." (In other words, according to this account, the men suggested that the women are just "living off" the men's income.) The authors then go on to note, "Although most men share their cash income from hunting and logging with their families, they also spend a substantial part of it on typical 'male' activities, such as drinking, smoking and other forms of entertainment in the district centres." In this particular case, it appears that the justification of periodically greater cash incomes, even from activities such as hunting and logging (which were not as strictly enforced at that time but were not allowed and thus brought in good amounts of cash), was used as a way to maintain the men's higher position in the gender hierarchy and ensure that women's hard work and totality of contributions were seen as relatively unimportant.

References

Adger, W. N. (1999). Social vulnerability to climate change and extremes in coastal Vietnam. *World Development*, 27(20), 249–269. https://doi.org/10.1016/S0305-750X(98)00136-3

Arhem, N., & Binh, N. T. T. (2006). *Road to progress? The socio-economic impact of the Ho Chi Minh highway on the indigenous population in the Central Truong Son region of Vietnam*. Presented for WWF Indochina. www.academia.edu/10639717/

Road_to_Progress_The_socio-economic_impact_of_the_Ho_Chi_Minh_Highway_on_ the_Indigenous_Population_in_the_Central_Truong_Son_Region_of_Vietnam

Bayrak, M. M., Tran, N. T., & Burgers, P. (2013). Restructuring space in the name of development: The socio-cultural impact of the forest land allocation program on the indigenous Co Tu people in Central Vietnam. *Journal of Political Ecology, 20*(1). https://doi.org/10.2458/v20i1.21745

Beckman, M. (2011). Converging and conflicting interests in adaptation to environmental change in Central Vietnam. *Climate and Development, 3*(1), 32–41. https://doi.org/10.3763/cdev.2010.0065

Ca Dy Commune People's Committee. (2012). *Annual socioeconomic reports from 2011–2012.* Vietnamese Version. London: World Bank.

Chaudhry, P., & Ruysschaert, G. (2007). *Climate change and human development in Vietnam.* Human Development Report Occasional Papers 1992–2007. HDOCP-2007-46. Human Development Report Office, United Nations Development Programme. https://EconPapers.repec.org/RePEc:hdr:hdocpa:hdocpa-2007-46.

Luu, H. (2007). *A contribution to Katu ethnography: A highland people of central Vietnam.* SANS Papers in Social Anthropology, 9. Gothenburg: University of Gothenburg.

World Bank. (2010). *World development report 2010: Development and climate change.* World Bank. http://hdl.handle.net/10986/4387

5   Summary of findings and possible ways forward

Analyzing combined impacts of climate, regulatory and social change

Climate change, together with regulatory policy and socioeconomic changes, has had profound impacts on the Co Tu community in Ca Dy Commune in Central Viet Nam. It is notable that elders and others interviewed for this study usually commented much more about the dramatic changes they have seen in the past two to three decades than about any previous changes – including those associated with the war years – which is one indication of the perceived strength of the impact of these combined forces on the local community.

It is important to note that a good deal of research on the impacts of climate change and on socioeconomic and regulatory changes on local communities has already been conducted to help people prepare for and better adapt to such changes. However, these studies have tended to focus on each topic individually rather than considering the *combined* impact of all these changes taken together.

It will be important as well to better understand the complicated ways in which climate change, together with other driving forces, can bring about changes in gender roles, relations and hierarchies. Gender-related considerations are often centrally important in determining whether adaptation can or cannot take place smoothly and effectively in communities that are suddenly faced not only with climate threats but also other profoundly destabilizing forces.

One such study of the complicated interaction of climate change and other factors is a recent analysis of food insecurity in the Peruvian Amazon (Zavaleta et al., 2018). This research looked at the relationship between *climatic* and *non-climatic drivers* of change that "reinforce climate change maladaptation trajectories" (i.e., tendencies toward "negative adaptations" when people are faced with new and potentially harmful combinations of natural and social forces). In the case studied by Zavaleta et al., these trajectories (responses) were influenced by such factors as national cash and food programs that targeted women and led to unanticipated problems, including men turning away from their traditional livelihood activities and women becoming overburdened with their new work and responsibilities.

Although the circumstances are different, the parallels with the present study are significant in terms of the analysis of multiple drivers of change that can lead to new challenges impacting women and men in different ways. This can, in turn, cause new problems and tensions and, in some cases, result in even greater vulnerability to climate threats and other sudden changes and requirements imposed on local communities.

Some of the greatest impacts of these profound changes are seen in the forms of *relationships between people* – relationships that would gradually, visibly and unintentionally be transformed as members of the community try to suddenly cope and adapt to the entirely new world created by these combined forces. In the case of the Co Tu community of Ca Dy Commune – as is true, to different degrees, of other ethnic minority communities in Viet Nam and minority and indigenous communities in many other parts of the world – their previously isolated and remote location, lack of resources and assets and lack of control over processes of change make this transition very challenging to the community.

One question, then, is: How can communities that are subjected to such destabilizing changes deal with and respond to these challenges as effectively as possible? We will approach this question by first presenting a brief summary of the findings of this research, followed by a discussion of how the findings of this case study relate to other studies regarding gender, ethnic minorities, climate, policy and socioeconomic change. We will then discuss different views on how communities – not only in Viet Nam but also in other similar contexts – might best respond to these complex challenges. Because each case is unique and highly contextual, no definitive "answers" can be given; still, we will try to draw out a few general implications based on the findings of this study and discussions in the relevant literature, drawing primarily (but not exclusively) from the experience of Viet Nam.

Summary of findings

Over the past two to three decades, the Co Tu community has been strongly affected by climate change, regulatory change and socioeconomic change. Figure 5.1 presents the collective impacts of changes on the Co Tu community's food and livelihood security as well as impacts on relationships within the household, specifically between Co Tu men and women of both the older and younger generations.

We have seen how under the combined impacts of changes in these three important areas (climate, regulation and society), the Co Tu community's traditional livelihood system was fundamentally and negatively affected. Along with livelihood security, food security was also under threat because of the damage induced by climate change and reduced available cultivated land area and consequent soil degradation, together with prohibitions on swidden (shifting) cultivation, hunting and logging and restrictions on the collection of non-timber forest products (NTFPs) and the trapping of wild animals from the forests.

Co Tu women have been concerned the most with the sudden loss of food security because they are responsible for providing food for their extended families. Women's workload has increased significantly while men's workload has decreased

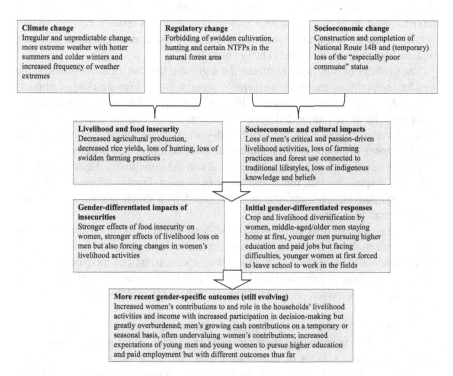

Figure 5.1 Summary of findings

considerably, particularly in the initial years after the coming of regulatory restrictions (instituted, in part, as an attempt to protect the natural environment). These restrictions were implemented in a way that suddenly deprived men of their traditional roles and occupations and, paradoxically, had significant negative effects as well, directly and indirectly, on the area's natural environment – including a wide range of trees, plants, soil and animals that became subject to deterioration, erosion and illegal exploitation once the villagers' shared system of forest use came to an end.

We have seen that gender roles and hierarchies have been changing over time but in a complicated way, particularly under the constantly changing pressures of climate threats, regulations and exposure to market forces. Both men and women have had difficulty coping with these sudden upheavals, resulting in new tensions, but also new possibilities for both women and men as they respond to these ongoing challenges.

Discussion of findings in relation to the literature

According to Momsen (2010), gender is "the socially constructed form of relations between women and men" and is the "acquired notion of masculinity and femininity by which women and men are identified" (p. 2). Gender definitions have focused on gender relations – including relations of relative equality or inequality – that are culturally constructed and reproduced among women and men and are also subject to change (Dankelman, 2010).

The findings of this study reinforce the argument that the more pronounced the gender inequalities, the more vulnerable women are to climate-change impacts (Arora-Jonsson, 2011; Neumayer & Plümper, 2007). In the Co Tu case discussed here, with its strict patriarchal system, the impacts of climate change and other forces have disrupted long-practiced, self-sufficient livelihood activities, resulting in greater exposure to the mainstream Kinh economy and society as well as imposing new forms of gendered livelihoods on the community. Because of their roles and responsibilities, the women have been most directly affected by climate and accompanying changes and are now faced with much greater stresses and burdens than in the past.

This has had serious implications for both women and men and for relations between them. In the Co Tu case, the earlier complementary relationships between socially designated women's and men's contributions have been entirely disrupted, leading to increased tensions, unequal burdens and new perceptions of "maladaptive" behavior, particularly those attributed to middle-aged and younger men.

The literature has shown that gender relations are contextually specific and vary in response to changing environmental and social circumstances, along with altering symbolic notions of masculinity and femininity (Moser, 1993). In this case, we see that both women and men can be seriously and negatively affected as their environmental and social circumstances change in uncontrollable ways, particularly if they are prevented from being able to meet social expectations and carry out their designated social responsibilities (Ashwin & Lytkina, 2004; Brittan, 2005; FAO & UNDP, 2002; Hoang & Yeoh, 2011; Kabeer, 2007; Salemink, 2003; Swinkels & Turk, 2006; Tran et al., 2006).

With respect to men in the community, this study found that men have been affected because they can no longer perform their traditionally designated roles due to the changes that have occurred in the past two to three decades. Under the impacts of imposed changes in livelihoods, reduced agricultural production due to climate and regulatory changes and increased impacts from the influence of the norms and values of the Kinh majority, a new set of priorities has come in based on the ability to make money rather than valuing the physical ability necessary to protect families from danger or secure agricultural and forest-based food and other necessities. Although the men are still struggling to find new roles while looking for new and better job opportunities, this study found that many men who were displaced from their traditional livelihoods initially spent a great deal of time drinking and being idle, contributing little to their families or communities.

Moreover, as noted in the previous chapters, in recent years the new value system has pushed men into a range of irregular activities for quick sources of cash. This appears to be a factor in less value being placed on women's *non-cash* contributions, no matter how important and extensive the women's contributions are to both their families and communities. These forces have also had the effect of reducing mutual support and social solidarity in the community and, in many cases, resulted in greater psychological and social stresses in the face of rapid change.

The study thus found that, given the symbolic notions of masculinity and the perception of men as breadwinners, men are also suffering from this process of change because they are losing their power and status, even though they are resisting this change. As Connell discussed in "Change among the gatekeepers: Men, masculinity and gender equality in the global arena" (2005), in a patriarchal society, when masculinity is associated with being the breadwinner and being "physically strong" – which still holds true in the Co Tu community – the increased role of women in the family and community often appears to challenge men's dominant position. This may result in a fear of women gaining dominance in a continuing *gender hierarchy*, with men in a lower status, rather than being seen as a movement toward *gender equality* (i.e., without a hierarchy in which either men or women dominate) with positive dimensions for men as well as women. (An example of the fear of being dominated by women appears to be reflected in the comment, quoted in Chapter 4, of a male FGD participant about a local woman who had done well in business and whose husband "obeys her very much" – a description that belittles the man in this cultural context and appears to place him in a lower position in the gender hierarchy.)

In this way, we can see that women are not the only ones experiencing the negative impacts of climate and accompanying changes. Men are also struggling to adapt to the new gender division of labor and decision-making processes, which can lead to tensions and conflicts that affect their own wellbeing and, in turn, the wellbeing of women, children and others around them.

The study further found that in recent years, rather than remain idle and be left out of a cash-based economy (with the falling social status that implies), many Co Tu men have begun to compromise on their vision of masculinity as strong and self-determining; in other words, they do not want to be someone else's employee, given their background as hunters and self-sufficient producers who freely determine when and where to carry out their livelihood activities. However, they generally cannot fulfill that ideal at the present time and instead have to take what they can find, as long as it brings in the cash that buys the items they now need and desire in the new money-based economy. The young men have also adopted this pattern, especially once they decide to get married.

The initial reluctance of middle-aged men to see themselves as "hired labor," and the continuing reluctance of many young men to take on jobs that require regular hours with little day-to-day flexibility, could be analyzed in terms of a move from an agrarian to a more industrial lifestyle. Difficulties in making this transition can be seen throughout history in different parts of the world. However, the added dimension here is the *gender hierarchy* that drives middle-aged women and younger women to do whatever is needed while men are allowed more freedom to choose their course of action – including how, when and where to work. Women, in contrast, do not have the freedom (e.g., to choose to remain idle or to drink, sleep all day or play video games) that men – and even young men – have in a rigid patriarchal system. This often results in their facing serious challenges that, so far at least, men and boys are not generally expected or required to face.

Nonetheless, this study suggests that this process of change has not been altogether negative for women in that the changes have presented new opportunities as seen in, for example, the women's increased self-confidence as they gained greater roles in the market.[1] The changes have by now opened up new opportunities for younger women as well. These new roles and activities are, at times, accompanied by greater decision-making power and recognition, both inside and outside the household, compared to their earlier status in the self-sufficient economy.

As noted previously, the women's rise in status has not gone unchallenged. In this case study, we see that even though women have had to take on greater livelihood and decision-making responsibilities, household members as well as the general population in the community continue to see men as heads of households and the primary decision-makers. Therefore, a persistent patriarchal system has hindered the women's progress toward gender equality, and once again we see that social change often takes place in complicated non-linear ways.

The future of gender roles and hierarchies in this community is not entirely clear, as livelihood patterns and the behavior of new generations continue to evolve. It appears that to some extent the middle-aged men in the community continue to fit Li's description of surplus populations in that their traditional livelihoods have been taken away, and they either have no access to stable work or are unprepared to engage in the work that is available on a regular basis (Li, 2010). Moreover, some members of the community have expressed a particular concern regarding the generation of young men who so far have been unable or unwilling to work in regular long-term employment and after marriage tend to follow their fathers into short-term employment for temporary and seasonal sources of cash-based income (choosing to work only as much as required to meet certain household expenses and their own consumption needs). They may be adopting new symbols of masculinity: Rather than the physical strength admired in the past, the new symbols may have more to do with earning and spending the new cash income, owning new commodities, using new "technologies" and behaving in ways that reinforce their sense of how men in the community are supposed to – and are allowed to – behave. They may not be doing this explicitly as a way to remain on top of the gender hierarchy but rather because they see it as "natural" for men to act in this way.

Whatever the motivation, it appears that at the present time both middle-aged men and married younger men have decided to contribute to livelihood activities in new ways but on their own terms. They do not appear to feel the pressure that is placed on women and girls to provide for their families in regular, consistent and ongoing ways, as reflected in both the paid and unpaid work that women are required to carry out. Again, these latter activities tend to be seen as a "normal" part of women's responsibilities in the community and are generally not valued in the way that men's earnings are valued. This may reflect the way most non-cash-earning activities are devalued in primarily cash-based economies throughout the world (often neither men nor women see these non-cash activities as centrally important, even though they are). However, this tendency to undervalue women's contributions may also be due, in the Co Tu case, to the higher status traditionally accorded to men's activities in this historically strict patriarchal context.

In spite of the persistence and even reassertion of local gender hierarchies, with men in higher positions, this study found that Co Tu women and girls are starting to question the rationality of the imbalance in this gender division of labor and in gendered decision-making processes. Their roles and voices in the family are generally much stronger than had been true in the past. In addition, awareness about gender equality and inequality has been raised in the community, examples from outside are being introduced, and the idea of women as submissive and inferior is no longer necessarily seen as being "natural" or inevitable.

Over time, the definitions of men's work and women's work, and women's and men's "proper" roles in their families and communities, may change as the community continues to be exposed to and participate increasingly in the mainstream economy and society. We see other cases of major changes in the gender division of labor, which indicate that such change in gender roles and hierarchies is possible, even when strongly resisted. For example, in a study of long-term changes in gender relations and economic roles in coal mining areas of the UK, when men lost their identities tied to their jobs and earlier lifestyle patterns, there arose paid job opportunities for women. The study found that the men's initial reactions to the change in the gender division of labor were almost universally antagonistic (McDowell & Massey, 1984). They considered women taking paid jobs to be an affront to men's masculinity and dignity. In the beginning, men stayed home but refused to help with housework while women took jobs outside their homes. Nonetheless, there was a process of gradual change with men progressively entering new forms of employment as well as increasingly helping with the reproductive sphere due to the influences of the work of community organizations, the media, the educational system, and other sources of change. Although the "working class culture" in the area was still said to be male-oriented (or male dominant) at the time of that study, traditional gender patterns had already been modified substantially. Similar changes have been noted in other case studies toward "more equal sharing of housework and child care" (Connell, 2005).

This may also happen in the Co Tu community as the area becomes less isolated, new ideas enter and new opportunities arise for both men and women. This transition, however, is not likely to be an easy one, particularly given the continuing pressures on the community tied to climate change and other rapid and unanticipated changes that have so disrupted the community, forcing it out of its earlier well-established practices and relationships in such a short period of time. The willingness of many young women and the reluctance of many young men to respond to new opportunities for study and full-time employment may signal that unequal gender hierarchies and strongly-held concepts of masculinity and femininity (and male entitlement) have not changed as much as might be expected, even under strong and increasing pressure. How these patterns change in the future will depend not only on factors outside the community's control but also on many factors under their control that can help reduce the community's vulnerability and may also be able to help reduce tensions and conflicts as well as move toward more equal gender relations in the future.

Implications for gender-related concerns: ethnic minorities and indigenous communities in previously isolated regions

This is a study of one small community in Central Viet Nam, but as indicated in Chapters 1 and 5, there are parallels in many other parts of the world as indigenous and ethnic minority communities in previously remote locations are affected by climate change in very negative ways. Climate change can have an immediate effect on their food and livelihood security, and because climate change often interacts with other powerful driving factors, it can force rapid change in local communities not only in a material sense but also in terms of social relations, including gender relations.

Too often, in indigenous communities that experience dislocation – and particularly in those that are required to move quickly toward the social and economic mainstream even before they can fully understand and deal effectively with the problems created by this transition – the results can involve a serious loss of self-esteem, frustration, in some cases self-harm (e.g., through substance abuse or depression), domestic conflict, gender-based violence and other types of "negative adaptations" due to the impact of these combined forces. When forced to compete when they are unprepared to do so, indigenous communities may also internalize views from some outsiders that they are "poor and backward," leading to further demoralization and resulting problems in the household and community.

Positive responses to these problems, in a context of ongoing climate threats and other sources of rapid change, are necessarily very contextual and will have to be based on a close understanding of local natural and social conditions. Nonetheless, in the discussion that follows, we will highlight some of the key lessons we think can be drawn from this study, with the hope that these suggestions would be useful in other similar contexts as well.

Decentralization of national policies to take local conditions into account

Policy-wise, the call for a "decentralization of national policies" is a theme that is very relevant to local communities – especially indigenous and ethnic minority communities – that are subject to ongoing climate threats and other serious pressures. It is often noted that in addition to the sociocultural problems created by centralized policies that lack modifications based on local conditions, a "one-size-fits-all" approach to land use, agricultural and forest-related regulations could in some contexts result in environmentally unsound practices that can harm both local communities and the local natural environment. In many cases, it may be that too many years have passed to further modify these policies or reverse them in order to address the problems these policies have created. However, in other contexts, a move toward the *decentralization* (similar to the "localization" or "contextualization") of land use, environmental and poverty reduction policies and programs to take both local social and environmental conditions – including the effects of climate change – carefully into account can be extremely important.

One idea that anthropologists and others have advanced would be to try to reinstate, to whatever extent possible, certain features of the earlier agricultural systems in indigenous or ethnic minority communities that they see as being more culturally appropriate and ecologically sound given the belief systems, traditional cultivation techniques and social solidarity of indigenous communities. An example of this, focusing on ethnic minorities in the Mekong Subregion, can be found in the writings of Arhem and colleagues (Arhem, 2009; Arhem & Binh, 2006; and in the context of Lao PDR, Arhem, 2010). Arhem and Binh's study (2006, pp. 13–15), for example, recommends that a subnational-level *department of indigenous affairs* be set up to help localize (contextualize) and adjust national policies to fit specific local conditions with plans developed through consultations with the communities, including supporting customary practices that are valuable from environmental and sociocultural points of view. They also suggest that department personnel bring in individuals who ideally would have expertise in social sciences, agronomy/forestry, biology and/or other relevant disciplines to conduct workshops and periodically meet with local communities and local officials to come up with the best solutions to problems as they arise.

This type of contextualization effort, involving modifications of national policies to meet local conditions, would be in line with the effort to support and promote indigenous cultures; it could also be tied to national education, media and other policies and programs designed to ensure respect for different cultural perspectives, particularly in contexts in which indigenous and ethnic minority groups are sometimes seen as simply "poor" and "backward." We should note that this type of institutional change (e.g., setting up a subnational department of indigenous affairs) could certainly be useful, but it is more likely to be carried out in countries that recognize the importance of indigenous communities. Moreover, this would need to be done carefully and in a way that avoids favoring particular individuals, families or groups within the local communities. As seen in the present study, local participation in decision-making is critical and will need to include *both women and men*, and ideally representatives of different age groups as well as other important social groupings, at some level in the decision-making processes in order to allow the localized programs and projects to achieve their goals in a sustainable way (i.e., with the entire community's participation and consent).

A more general question to be addressed is the degree to which indigenous and ethnic minority communities should try to integrate with the national mainstream as a development strategy. Anthropologists and other social scientists often emphasize the need to strengthen *self-reliance and resilience* in local communities and avoid uncontrolled integration into the national mainstream when the community is not fully prepared for this. However, by stressing the importance of maintaining a high degree of self-sufficiency, there is also a potential for observers to underestimate the degree to which members of local communities now actually desire integration with the national mainstream. Moreover, climate change, population growth and other considerations may not allow for a return to practices closer to what was there before (e.g., livelihood and cultural practices associated with the earlier swidden cultivation system).

In contrast, economists tend to see education and a quick and closer *integration into the mainstream economy* as a key development strategy for poor communities (e.g., through value chain integration). However, these proposals are often made without a careful consideration of other factors, such as unequal access to contacts and information, that put ethnic minorities or indigenous groups at a serious disadvantage when dealing with mainstream agents and institutions. These inequalities would strongly caution against this type of quick integration (which, again, tends to be another "one-size-fits-all" recommendation that is found when rural development is approached without adequately contextualization based on gender relations, minority status and other key considerations).

For many indigenous communities, the choice may very well be to tie in more closely to the mainstream rather than revert entirely to earlier practices but with some degree of control over the direction of change. Some members of the community may also try to maintain traditional practices selectively, without fully returning to the previous patterns that placed them lower in local social hierarchies, including gender hierarchies. As an example, a study by Vu et al. (2011) described efforts to revive traditional knowledge and systems of forest protection, but by combining traditional practices with new forms of women's involvement in decision-making, promoting gender equality in the inheritance of property and ensuring increased local control over forests and community development. Without these specific types of initiatives to advance the more vulnerable segments of the local population, such as initiatives regarding women's access to land, it is by no means certain that those lower in traditional social hierarchies, and members of the younger generation who now have been exposed to altogether different values, would actually want to return to the "way things were" (for example, to return to the earlier ways of doing things, including customary law that does not in any way ensure the rights of women to even make decisions regarding their own lives).

This is a complicated question, in part because aspirations are likely to differ depending on the experiences of different social groupings, such as age groups. For example, regarding gender relations and hierarchies, an older generation may have been used to what they see as the *complementary* roles of men and women reflecting the gender roles and relations that they grew up with, rather than sharing the perception that women are being treated unfairly. However, some women – particularly younger women – may have different desires for their place in society and conceptions of what is just and fair, given the changes that have already occurred, their experiences in recent years (including developments such as women being overburdened), and their exposure to different views regarding gender roles, relations and hierarchies.

In cases where a return to a more self-sufficient and relatively isolated status is not possible or desired, most observers would agree that depending on the local and national context, institutions such as government agencies and NGOs are likely to be needed to provide the educational, health, training and other facilities that will help local communities deal with mainstream requirements for education, language, work experience, personal connections, capital and other factors that local community members may lack, thus putting them at a serious disadvantage (making them more subject to exploitation, not getting what they deserve, being excluded from opportunities and other difficulties). These facilities are often weak or not available in remote areas, but strengthening this infrastructure and preparing members of the local community – women and men – to deal with more mainstream contexts will be of key importance, no matter what particular development strategy is adopted. As part of this, the decentralization (and localization/contextualization) of national policies will serve as a crucial element in making these strategies work toward both social and environmental goals.

Viet Nam's national policies, contextualization efforts and ethnic minorities

One example of a *centralized* approach that impacts ethnic minority communities in Viet Nam is the National Target Program on New Rural Development (NTP-NRD), one of the largest national-level programs in Viet Nam. The NTP-NRD was approved by the government on June 4, 2010, and covers the decade 2010–2020.

This program is recognized as one of the greatest forces shaping Viet Nam's pattern of rural development over the last ten years as it is designed to develop rural areas with improvements in many key sectors: infrastructure, economic improvement, poverty reduction, social security, production structure, health care, education, cultural and spiritual life and many other areas of rural life. Nine thousand communes across the country have attempted to meet the new rural development standard of success, involving 19 criteria to determine the degree of improvement.

However, in late 2015 to early 2016, IFAD and the World Bank identified one of the remaining shortcomings of the program in a joint assessment, which found that the program was not adequately contextualized. Because of this, in some cases the program had neglected local development priorities and needs (IFAD, 2016).

In 2019, the Government of Viet Nam, together with other stakeholders, reviewed the program's history of implementation since 2010.[2] They found that although there have been notable successes, mostly in terms of economic improvements as well as improvements in infrastructure, environment and living conditions, several discussions have centered around the need to revise the set of 19 criteria in order to meet the needs and circumstances of local communities.

The findings of this study of Ca Dy Commune support the argument put forward in the IFAD-World Bank assessment regarding the need to decentralize and contextualize the program. As part of this, we find that education- and training-related initiatives need to be tailored to the needs and circumstances of local communities in general and of ethnic minority groups in particular. It is important that cultural characteristics, employment availability and local people's goals as well as advantages and disadvantages be taken carefully into account so that key national programs can make the best use of financial and human resources while providing local communities with sustainable livelihoods and meeting other critical *material and non-material needs* – again taking environmental, social and economic requirements as central to all decisions.

In fact, the next phase of this program is believed to focus on sustainable development with special attention to cultural and social development and priority given to especially disadvantaged communities – which would include ethnic minority communities, such as the Co Tu of Ca Dy Commune. Still, we would further argue that within this framework, the ethnic minorities' cultural identities and priorities, social inclusion, and gender equality as well as environmental and other goals should be considered explicitly as key principles.

Regarding initiatives to address livelihood needs that might have a positive social impact, we should note that one of the main findings of the present study was that a key strength of the Co Tu community in the past was their social cohesion and solidarity. After they were affected by and required to move into the cash economy, they have engaged less in community and group activities and function increasingly as individuals. However, as individuals, they need to compete with the Kinh majority in many employment markets, and for this they still face considerable disadvantages.

Educational and training programs would, therefore, ideally be linked to a more *comprehensive livelihood plan* that ensures trainees will not only gain new knowledge and skills but will also be able to use those skills and build on them as needed – particularly in association with others, rather than as individuals competing directly with better-off and better-connected members of the ethnic majority. It will be tremendously frustrating if trainees do not have access to job opportunities that will employ local ethnic minorities, rather than the jobs preferentially going to ethnic majorities; it will also be frustrating if developments such as changes in market demand result in their new skills no longer being useful or if the new livelihoods prove to be environmentally unsound. Under rapidly changing conditions, they will need the *flexible knowledge* to expand their range of skills or modify the ways in which their skills are used and learn to find the contacts and build the relationships needed to adjust to rapidly changing social, environmental and market conditions. If the trainings offered are too short-term in vision and too narrowly focused (inflexible) to deal with change, the failure of these initiatives could further erode the self-confidence of ethnic minorities who already face significant disadvantages.

A more comprehensive set of initiatives for ethnic minorities should start with social impact assessments, considering changing gender dynamics as well as sociocultural requirements (responding to problems such as the ones detailed in the present study), along with environmental and economic assessments, health-related initiatives and other complementary efforts as needed to make livelihood goals both attainable and sustainable. These initial assessments will also help identify the trainings required to move toward the community's stated goals, which should include and reflect the goals of different groups within the community.

Training programs associated with community-focused initiatives should also involve *periodic follow-ups* from local or national agencies, given that continuing support will likely increase the sustainability of these efforts. (This is true even in cases in which few additional resources are required; the social benefits of having local and national-level institutions provide *continuing guidance and support* are often a key factor in sustainability.) If done with community involvement and consensus, these initiatives should also help strengthen the community's sense of solidarity, which in this study was referred to as an important part of the "old happy days." The educational and training programs and the overall development plan should emphasize the integrity of ethnic minority groups and help instill a sense of pride and self-confidence, rather than try to make ethnic minorities simply conform (less than successfully, at least initially) to patterns associated with the Kinh ethnic majority.

In fact, there is currently some movement in Viet Nam toward planning that takes into account ethnic minorities' specific needs and circumstances: At the present time, the Government of Viet Nam is considering a proposal (not yet approved) for another National Target Program (NTP) known as the National Target Program on Socio-economic Development in the ethnic minorities and mountainous areas for the period 2021–2030.[3] Even though some have argued that there is no need for a separate program, given that there are already two NTPs that can be applied in areas with ethnic minorities (the previously mentioned NTP-NRD and the NTP on Sustainable Poverty Reduction), we would argue that this new program, if approved, could provide an opportunity to overcome challenges associated with the two existing NTPs. This is due to the fact that although the other two NTPs may try to tailor rural development criteria to local contexts, they still do not adequately address the special geographic, environmental, social and cultural conditions, needs and priorities of Viet Nam's ethnic minorities in particular.

As part of this effort, the program was developed by the government to submit to the 14th National Assembly at the ninth session in 2020. At this stage, the program sets high targets, such as aiming to provide training to more than 50% of the laboring population; securing stable-income jobs for more than 80% of the population 15 years and older; establishing kindergartens, primary and secondary schools in 100% of the communes in these areas, with more than 80% of communes having school facilities that meet the new rural standards; ensuring literacy for 94% of working-age members of ethnic minorities; and training over 90% of the cadres and civil servants of communes with knowledge to ensure an increased understanding of ethnic minorities. The program proposes focusing on key mechanisms, such as residential planning combined with infrastructural development at the commune and village levels; investing in ethnic minorities through human resource development; livelihood creation and business start-ups, including starting businesses based on cultural diversity; and taking advantage of regional benefits and attractions, including cultural uniqueness and environmental diversity. Notably, the program prioritizes gender equality and solving urgent issues for women and children, and it is expected to open up ways for ethnic minority women to be empowered with effective access to and participation in new opportunities.

However, even though the proposal discusses the importance of cultural and natural uniqueness and the need for additional knowledge regarding ethnic minorities, even at very local levels, we would suggest that an *understanding of and respect for ethnic minorities* needs to be put at the center as one of the key principles of the program if it is approved – rather than, for example, focusing on cultural and environmental uniqueness primarily to generate employment and commercial opportunities in ways that may not be sustainable. One of the main challenges facing a national program for socioeconomic development is that economic targets tend to be emphasized over social ones, and it is difficult to incorporate *social* (including gender) and *environmental* priorities within this framework. Moreover, limited human resources and capacity of government officials generally work against detailed and decentralized assessments of needs and impacts on local ethnic minority communities. As an indication of the difficulty of breaking away from a centralized top-down approach, we should note that a locally driven and participatory approach to development planning has been recommended for the NTP-NRD since 2016, but this still has not been integrated into the program due to the already high costs and limited capacity of government agencies doing the planning and implementing of the program.

In spite of these difficulties, an NTP dealing with ethnic minority groups – if it were to succeed in becoming more contextualized and locally driven rather than being top-down, centralized and primarily oriented toward standard economic and commercial goals – could provide a good example of the types of efforts many have suggested are necessary in order for ethnic minorities and remote and indigenous communities to overcome the significant barriers they face. As our research findings indicate, not only do material needs have to be addressed in terms of food and livelihood security (which, of course, are critical for local communities), but also the *social barriers* ethnic minorities face and the *social impacts* of rapid change need to be addressed if improvements, rather than destabilizing and unhealthy outcomes, are to be the future for ethnic minority communities.

Promoting social cohesion and gender equality through sustainable local institutions

As noted earlier, previously remote indigenous and ethnic minority communities are often characterized as having placed great emphasis on social solidarity and cohesion and as having taken care to avoid the emergence of social differences that might pull the community apart. However, with their entrance into a cash economy, differences based on income, wealth and growing ties to the outside world will tend to erode the community's interdependence and sense of social cohesion over time. Moreover, as the focus shifts to individual and household-based striving and accumulation of assets, and away from shared achievements and experiences on the clan, village or larger-community level, both interpersonal conflicts and social isolation are likely to increase.

Ideally, interventions would use democratic means to strengthen social cohesion and promote community-based groups, organizations and other institutions and structures that help build social solidarity without favoring specific social groups, families or individuals over others. Even in rapidly changing and newly cash-based contexts, local interdependence and solidarity can be promoted; moreover, creating *ongoing institutions* that are locally appreciated and valued will increase the likelihood that social cohesion will be sustained in the local community.

In some contexts, *groups, organizations and networks* as well as other local institutions and structures – whether tied to livelihoods, health, working with government agencies on needed facilities or other objectives – can be beneficial in bringing community members together and promoting social cohesion. As noted earlier, in a patriarchal society, women in particular may become isolated, and promoting groups and organizations that allow women to meet, work together and have greater influence can be very helpful. However, it is important that these groups (e.g., those organized toward livelihood objectives) should not be perceived as benefiting group leaders or other specific individuals within the group for their own benefit and to the long-term detriment of other members. These groups, organizations and networks should also be clearly inclusive and work to benefit the community as a whole.[4] In vulnerable ethnic minority communities, well-regarded and well-functioning local institutions and structures can serve as crucial sources of self-identity, pride and strength, particularly when dealing with larger mainstream institutions.

Gender assessments and trainings and the importance of new knowledge

In a context of rapid change, and given that gender inequalities can worsen or improve depending on the direction of change, we would suggest that, along with interventions to help communities cope with serious social, economic and environmental challenges, periodic *gender assessments* need to be included in analyses prior to, during and after interventions, such as those outlined earlier, have been implemented. This assessment should provide an analysis of women's and men's changing roles and responsibilities and recognize the difficulties and disadvantages men *and* women would need to overcome as the world changes so suddenly around them.

For example, in the case of poverty reduction and social protection–related interventions, gender has to be taken clearly into consideration, particularly because tensions often arise if either men or women are the focus of the interventions in a way that increases conflict. We sometimes find that in patriarchal systems men can be resentful if women rather than men are the direct targets of such programs, with serious consequences if men's requests are not met. (This approach, targeting women in particular, has been followed in recent years in places where men are seen as taking a large part of the benefits for their own consumption, whereas women have been found to return benefits to the entire family.) Again, to avoid this type of serious conflict, programs and projects need to be designed with a clear knowledge of local gender relations and hierarchies and ensure that both men and women are "on board" as interventions are planned and implemented, with intermittent reassessments to see if the intended beneficiaries are actually receiving the benefits without creating new sources of conflict. Periodic evaluations using detailed sex-disaggregated data, such as those concerned with patterns of income inequality, poverty, health (emotional and physical), educational levels and the like, can be incorporated into gender-assessment efforts.

Regarding livelihood-related interventions, proposals for indigenous and ethnic minority communities often focus on educational systems and training programs that allow greater access to jobs and livelihoods, and in the case of communities in previously remote areas, these programs often involve technical trainings on agricultural production methods. However, each of these interventions has important gender implications: For example, as mentioned in the previous chapter, even though agricultural extension services have, in many parts of the world, traditionally focused on men, the actual cultivation may be undertaken by women – or, in some cases, the opposite is true.[5] In addition, access to and ownership of land and other assets are often highly gendered, and whenever interventions are proposed, the gender dimensions and impact on gender relations should be taken into account carefully and systematically, particularly if inequitable conditions of knowledge, access and ownership are to be overcome through targeted interventions over time.[6]

Moreover, as emphasized throughout this study, finding more *sustainable* livelihood activities for both men and women will be central to these interventions, including capacity-building that allows a broad and diversified range of potential livelihood activities, given that climate, market demand and other influential forces keep changing. Moreover, as noted earlier, bringing in scientific and technical knowledge as needed (e.g., pertaining to climate science, veterinary, agroforestry, hydrology, craft-related and other relevant disciplines) will increase the status and appeal of new livelihood activities, including in agriculture; for example, the incorporation of new scientific and technical knowledge will help raise the status of both women and men farmers in contexts that have devalued such activities, and add to their self-confidence and positive sense of their identities as skilled farmers, lay veterinarians and the like. (As one respondent put it, "I think we are farmers. It would be better if we were good at farming.") Local communities will also need to know more about how to deal with the surrounding market economy, including ways to find new markets, create ongoing personal relationships with traders and ensure fair and decent prices associated with their products, whether agricultural or non-agricultural. Above all, ways to anticipate climate threats and *strategies to deal with a changing climate* will be fundamental to communities that depend so closely on the natural environment that surrounds them.

For these reasons, and as noted previously, more educational programs and trainings related to sustainable livelihoods should be designed to include periodic follow-ups, keeping in mind the skills and desires of different genders and age groups. These initiatives would ideally also incorporate and highlight the knowledge and perspectives of older generations (elders) not only for their intrinsic value (e.g., as part of "technology blending" that includes the use of both traditional and modern science-based techniques and an understanding of the local natural environment) but also to provide cultural continuity, dignity and a sense of being part of an ongoing community with its own self-identity, integrity and sources of knowledge.

Combining "newer" and "older" sources of knowledge about livelihoods and the natural environment, and the successful application of that knowledge, should not only increase women's and men's statuses in the family and community but will also strengthen their own perception of themselves. In many contexts, we find that women's and men's sense of self-worth is closely tied to an increase in knowledge, capabilities and ability to contribute to their families and communities, which comes from their own self-perception as well as the recognition gained from others around them (Doneys et al., 2019).[7]

With respect to avoiding increased tensions, including the possibility of increased domestic conflict as gender hierarchies change, gender-related problems should be identified and addressed through periodic responses and monitoring by local institutions, government agencies, NGOs and other organizations, depending on the context. *Gender-awareness trainings*, for both women and men, will be important as well, but they are not likely to be effective if they are not "localized" and followed up in a systematic way. In addition, trainings in *leadership, communication skills, anti-violence measures* and other skills that will help improve women's and girls' as well as men's and boys' positions in their families and communities can be an important part of these efforts, depending on what local needs are identified.

Equal access for both women and men to public activities, such as those mentioned earlier, will have the additional benefit of helping women improve their social networks and reduce their isolation, a problem that often occurs where women are much lower in gender hierarchies. These measures should, in turn, help create more gender-equitable relationships and greater involvement in decision-making within the household and community, including better control for women over family property and other assets, without creating intra-household and intra-community conflict.

Initiatives of this type, if conducted in appropriate and culturally sensitive ways, can contribute to increasing the community's and family's recognition of women's roles and contributions and can help men transition to new roles. Such initiatives, including specific trainings as needed, ideally would involve the participation of respected community members to give a sense of their importance and, if possible, should be linked to health, education and other programs that *continue over time*, rather than involving simple "informational trainings" that are brief and occur only once or very occasionally.

Finally, given that in rapidly changing contexts different age groups may have very different conceptions regarding gender roles, relations and hierarchies, we should note that it is useful to take age, gender and other sources of social differentiation into the design of specific trainings and interventions. For example, several countries have employed "envisioning" activities that have been conducted with older, middle-aged and younger women and men and are likely to indicate very different conceptions on the part of each age group regarding their own long-term goals. These conceptions can then be incorporated into local planning for needed trainings and initiatives as community members participate in the development and modification of proposed interventions.

A closing thought

As we have seen, the story of Ca Dy Commune is not a simple one. When looking at gender relations in this case, it is very easy to fault men for not seeming to see or value the work that women have been doing, for trying to keep their wives away from training opportunities and public activities (because their wives "might meet other men," among other reasons) or for assuming that men, but not women, are entitled to remain idle or take irregular jobs if they so desire. These examples of male entitlement can be seen in many contexts internationally, and these patterns certainly tend to emerge in the context of strict gender hierarchies and patriarchal systems. On top of this, we also see cases throughout the world of increases in substance abuse, intra- or inter-community conflict and domestic and gender-based violence often associated with men who cannot respond well as the world suddenly changes around them in ways that are out of their control. The question then becomes, why do we see these patterns so frequently, and what might be done to resolve the complex problems that come with too great pressure and too rapid change?

Here, we would like to highlight the extremely destabilizing combination of factors that can challenge the core values, social solidarity, traditional relationships and patterns of respect that were hallmarks of the earlier economy and society in Ca Dy Commune. We can blame the old patterns as being harmful in many ways, but in a context in which relationships were previously seen as natural and complementary, we can still recognize that when men suddenly lose their key social roles, when women suddenly are overworked, overburdened and undervalued and when elders and traditional leaders are no longer listened to, demoralization can set in, and new sources of conflict will arise.

However, with access to knowledge, ongoing support and well-regarded and well-functioning local structures that are a source of pride and social cohesion, we would argue that ethnic minority and indigenous communities are more likely to gain the resilience to respond effectively to new, irreversible conditions. The climate threats will no doubt continue to worsen (at least until fundamental changes are made), the environment may continue to suffer and the local communities will most likely continue to integrate even more closely with and be influenced by the relatively individualistic and consumption-oriented mainstream economies and societies that surround them.

Interventions such as understanding clearly the direction of change; trying to prepare adequately for these changes and reduce vulnerability; strengthening a sense of the value of the entire community, including both women and men; and reducing sources of conflict between genders, communities and generations through better mutual understanding and cooperative efforts will be needed for better outcomes, rather than putting blame on men, on "one-size-fits-all" (centralized and not contextualized) policies or on any other single factor without understanding what is driving these decisions and behaviors. As emphasized repeatedly, local men as well as women – including the young women and men of the next generations – will be central to these efforts, and respected community members as well as national and other policymakers will have to be engaged in order to make these efforts work.

Of course, each case will be unique, and these types of proposals will not apply to all such communities. Still, the fundamental principles of interventions to increase technical skills, a blending of "old" and "new" knowledge, mutual understanding and cooperation, social equality, social solidarity and self-confidence with a strengthened sense of self-worth for women, men and the community as a whole, will be crucially important in all cases. At a minimum, this is a good place to start.

Notes

1 We have seen this in many other studies of gender and climate change: For example, a study in Tanzania indicates that, as a result of climate change, women's workload increased substantially when they were forced to switch to different crops. However, as in the Co Tu case, this also led to the benefit of higher independent earnings for women, again with both positive and negative effects resulting from their need to shift production in order to maintain their livelihoods under difficult climate change–related conditions (Nelson & Stathers, 2009).
2 Information from one of the review conferences on the NTP-NRD (July 2019) and documents (in Vietnamese) distributed during the conference can be found at this government website: http://nongthonmoi.gov.vn/FileUpload/2019-07/BzSdUUqQLUKOyIxoK%E1%BB%B6%20Y%E1%BA%BEU%20PHI%C3%8AN%20TO%C3%80N%20TH%E1%BB%82.pdf.
3 An initial announcement about this program, as it is being currently discussed, can be found at: www.dangcongsan.vn/tu-tuong-van-hoa/can-mot-chuong-trinh-muc-tieu-quoc-gia-ben-vung-vung-dan-toc-thieu-so-531915.html. (This is a journal article in Vietnamese from a government website regarding the need for an NTP for ethnic minority areas.)
4 More will be said about potential benefits and problems associated with the creation of groups, organizations and networks as well as other local institutions and structures in the authors' upcoming work on women's empowerment and security projects in four Mekong countries (Cambodia, Lao PDR, Myanmar and Viet Nam).
5 Two examples illustrating how gender concerns need to be taken into account when planning trainings can be given by the lead author: On one hand, she found in Hoa Dong Commune (Tay Hoa District, Phu Yen Province) that due to gender stereotypes of men being heads of households, agricultural extension, market and business-development trainings have often ended up with men far outnumbering female farmers in spite of the fact that both take part equally in agricultural work (from fieldwork conducted in 2019). However, gender stereotypes of this sort need not always prevail: For example, in a 2016 case study she conducted regarding a rural community near Hanoi (Thuong Phuc

Village, Dong Phu Commune, Chuong My District), women accounted for more than 70% of participants in most of the agricultural extension trainings, given that in this case many men often took off-farm jobs and left agricultural work primarily in the hands of women. Care must therefore be taken to include women and men according to the reality of their current work and their aspirations rather than allowing gender stereotypes to go unquestioned.

6 We should note the existence of local matriarchal/matrilineal as well as patriarchal/ patrilineal social systems in parts of Viet Nam, along with other South and Southeast Asian countries and elsewhere in the world, that are entirely counter to assumptions about ownership and access always being passed down through the male line. In some contexts, there are also conflicts between rules of inheritance based on state law vs. customary law, with mixed results. In general, it is important to keep in mind that there is actually a great deal of variation in the region tied to gender roles, relations and hierarchies, even if it is not always apparent at first glance; for this reason, careful gender assessments, rather than simply assuming that local communities all employ patterns associated with patriarchal social systems, are crucial.

7 As one example of an initiative for forest-dependent ethnic minority communities, we can come back to ideas that focus on using new knowledge for the development of forests, including forest management and climate-change monitoring, which will also aim to provide employment to the local population and help with the expansion of the forest cover. We would suggest that engaging indigenous and ethnic minority communities in efforts to restore fragile but crucial environments will be particularly effective if younger men and women are recruited as key agents in these reforestation projects; however, in cases such as the Co Tu of Ca Dy Commune, we have seen that the local population will not engage in reforestation efforts unless the programs are designed in a way that is appropriate to their specific needs, circumstances and priorities. Taking the concerns and desires of local populations into account is particularly important in the case of low-income ethnic minority communities that are likely to have needs and priorities that are quite different from ethnic majorities – and we have seen that *not* recognizing their needs, circumstances and priorities very often leads to the failure of social- and environment-related projects and programs. In addition, recognizing the value of traditional knowledge held by local women and men with respect to forest-based natural systems will reinforce the sense of self-worth for communities that regard forests as a source of many medicinal, cultural, spiritual and other benefits and not just commercial ones.

References

Arhem, N. (2009). *In the sacred forest: Landscape, livelihood and spirit beliefs among the Katu of Vietnam.* SANS Papers in Social Anthropology, 10. Gothenburg: University of Gothenburg.

Arhem, N. (2010). *Traditional ecological knowledge of ethnic groups in SUFORD AF production forest areas: A rapid assessment.* For Lao People's Democratic Republic Ministry of Agriculture and Forestry, Department of Forestry, Sustainable Forestry for Rural Development Project – Additional Financing (SUFORD – AF). www.academia. edu/10677446/Traditional_Ecological_Knowledge_of_Ethnic_Groups_in_SUFORD_ AF_Production_Forest_Areas_A_Rapid_Assessment

Arhem, N., & Binh, N. T. T. (2006). *Road to progress? The socio-economic impact of the Ho Chi Minh highway on the indigenous population in the Central Truong Son region of Vietnam.* Presented for WWF Indochina. www.academia.edu/10639717/Road_to_ Progress_The_socio-economic_impact_of_the_Ho_Chi_Minh_Highway_on_the_ Indigenous_Population_in_the_Central_Truong_Son_Region_of_Vietnam

Arora-Jonsson, S. (2011). Virtue and vulnerability: Discourses on women, gender and climate change. *Global Environmental Change, 21*(2), 744–751. https://doi.org/10.1016/j.gloenvcha.2011.01.005

Ashwin, S., & Lytkina, T. (2004). Men in crisis in Russia: The role of domestic marginalization. *Gender & Society, 18*(2), 189–206. https://doi.org/10.1177/0891243203261263

Brittan, A. (2005). *Masculinity and power*. Oxford: Basil Blackwell.

Connell, R. W. (2005). Change among the gatekeepers: Men, masculinities, and gender equality in the global arena. *Signs: Journal of Women in Culture and Society, 30*(3), 1801–1825. www.journals.uchicago.edu/doi/10.1086/427525

Dankelman, I. (2010). Introduction: Exploring gender, environment and climate change. In I. Dankelman (Ed.), *Gender and climate change: An introduction*. London: Routledge.

Doneys, P., Doane, D. L., & Norm, S. (2019). Seeing empowerment as relational: Lessons from women participating in development projects in Cambodia. *Development in Practice*. https://doi.org/10.1080/09614524.2019.1678570

FAO, & UNDP. (2002). *Gender differences in the transitional economy of Vietnam: Key gender findings. Second Vietnam Living Standards Survey, 1997–98*. FAO. www.fao.org/3/AC685E/ac685e00.htm

Hoang, L. A., & Yeoh, B. S. A. (2011). Breadwinning wives and "left behind" husbands: Men and masculinities in the Vietnamese transnational family. *Gender and Society, 25*(6), 717–739. https://doi.org/10.1177/0891243211430636

IFAD. (2016, August). *Review of experience of the national target program for new rural development: Vietnam.* www.ifad.org/documents/38714170/40253395/Policy+case+study+Viet+Nam+%E2%80%93+Review+of+experience+of+the+National+Target+Program+for+new+rural+development.pdf/bf68c25c-7489-4b06-b5ca-3095c9e4ab0f

Kabeer, N. (2007). *Marriage, motherhood and masculinity in the global economy: Reconfigurations of personal and economic life*. IDS Working Paper No. 290. Institute of Development Studies. University of Sussex. www.ids.ac.uk/files/dmfile/Wp290.pdf

Li, T. M. (2010). To make live or let die? Rural dispossession and the protection of surplus populations. *Antipode, 41*(s1), 66–93. https://doi.org/10.1111/j.1467-8330.2009.00717.x

McDowell, L., & Massey, D. (1984). A woman's place? In D. Massey & J. Allen (Eds.), *Geography matters!: A reader* (pp. 128–147). Cambridge: Cambridge University Press.

Momsen, J. (2010). *Gender and development*. London: Routledge.

Moser, C. (1993). *Gender planning and development: Theory, practice and training*. London: Routledge.

Nelson, V., & Stathers, T. (2009). Resilience, power, culture and climate: A case study from semi-arid Tanzania and new research directions. *Gender and Development, 17*(1), 81–94. https://doi.org/10.1080/13552070802696946

Neumayer, E., & Plümper, T. (2007). The gendered nature of natural disasters: The impact of catastrophic events on the gender gap in life expectancy, 1981–2002. *Annals of the Association of American Geographers, 97*(3), 551–566. https://doi.org/10.1111/j.1467-8306.2007.00563.x

Salemink, O. (2003). *The ethnography of Vietnam's central highlanders: A historical contextualization, 1850–1900*. London: Routledge.

Swinkels, R., & Turk, C. (2006, September 28). *Explaining ethnic minority poverty in Vietnam: A summary of recent trends and current challenges*. Draft Background Paper for CEM/MPI Meeting on Ethnic Minority Poverty. Washington, DC: World Bank.

Tran, T. M. D., Hoang, X. D., & Do, H. (2006). *Định kiến và Phân biệt đối xử theo Giới [Prejudice and gender discrimination]*. Hànội: Nhà xuất bản đại học quốc gia Hànội.

Vu, T. H., Nguyen, T. T., Nguyen, X. G., & Vu, H. X. (2011). Ethnic minority women in traditional forest management at Binh Son Village, Thai Nguyen Province, Vietnam. In W. V. Alangui, G. Subido, & R. Tinda-an (Eds.), *Indigenous women, climate change and forests*. London: Tebtebba Foundation, CERDA.

Zavaleta, C., Berrang-Ford, L., Ford, J., Llanos-Cuentas, A., Carcamo, C., Ross, N. A., Lancha, G., Sherman, M., & Harper, S. (2018). Multiple non-climatic drivers of food insecurity reinforce climate change maladaptation trajectories among Peruvian indigenous Shawi in the Amazon. *PLoS One*. https://doi.org/10.1371/journal.pone.0205714

Appendix

Photo 1 Cultural-sports festival: restoring a traditional Co Tu festival (in a location outside, but similar to, the study area)

Photo 2 A Co Tu elder sharing stories of his early days

Photo 3 A few of the Co Tu youth and the lead author in an informal discussion

Photo 4 Co Tu girls collecting firewood

Index

Printed in the United States
By Bookmasters